NANOSTRUCTURES IN ENERGY GENERATION, TRANSMISSION AND STORAGE

Edited by **Yanina Fedorenko**

Nanostructures in Energy Generation, Transmission and Storage
http://dx.doi.org/10.5772/intechopen.74917
Edited by Yanina Fedorenko

Contributors

Daejoong Kim, Majid Shahbabaei, Tatyana Shabatina, Radenka Maric, Haoran Yu, Corneliu Doroftei, José García-Antón, Bianca Lucas-Granados, Rita Sanchez-Tovar, Ramón M. Fernández-Domene, Toshio Takeuchi, Yoshiji Horikoshil, Petru-Andrei Bertea, Liliana Rozemarie Manea, Yanina Fedorenko

First published in London, United Kingdom, 2019 by IntechOpen
IntechOpen is the global imprint of INTECHOPEN LIMITED, registered in England and Wales, registration number: 11086078, The Shard, 25th floor, 32 London Bridge Street
London, SE19SG – United Kingdom
Printed in Croatia

British Library Cataloguing-in-Publication Data
A catalogue record for this book is available from the British Library

Additional hard copies can be obtained from orders@intechopen.com

Nanostructures in Energy Generation, Transmission and Storage, Edited by Yanina Fedorenko
p. cm.
Print ISBN 978-1-78985-739-9
Online ISBN 978-1-78985-740-5

Meet the editor

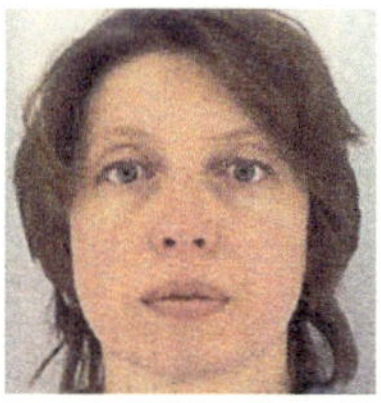

Dr. Y. Fedorenko graduated from Saratov State University, Russia, in 1998. Since 2001 she works as a post-doctoral researcher and scientist for companies and universities in Belgium, the UK and Finland. Her scientific interests include thin film synthesis, semiconductor processing and investigation of optical, structural and electronic properties of thin films and nanostructures.

Contents

Preface

I hope that this book, composed as a volume of self-contained research articles, appeals to a wide readership that is interested in phenomena occurring at the nanoscale in a variety of seemingly disparate materials systems.

New water treatment technologies are indispensable in the supply of safe water for domestic and industrial needs. Desalination is a method of choice to address the demands of water consumption. Membrane processes in water purification technologies are versatile and economically beneficial for drinking water production, desalination, water reuse, and wastewater treatments.

In research written by M. Shahbabaei and D. Kim, the authors consider water transport mechanisms in multilayer nanoporous graphene, which features the asymmetric hourglass nanopore structure of aquaporin protein water channels that facilitates fast water transport and results in an improved desalination performance.

Because of the unique capability to tune nanostructure geometry, surface morphology and charge, nanomaterials are exploited in virtually any kind of sensor devices because of the ability to tailor the sensing functions. The chemical composition, surface conditions, and crystal structure of nanomaterials are critical in adjusting transduction mechanisms. In this book, cost-effective approaches to the fabrication of nanomaterials are represented by the electrospinning technique, which is explored to prepare electrospun nanostructures. Electrospinning is a versatile method that, in addition to polymers, is suitable to synthesize other materials such as ceramics and carbon in a nanofiber form.

In the following chapter, L. R. Manea and A.-P. Bertea review electrospinning techniques for producing polymer nanofibers and elucidate a variety of applications of electrospun nanofibers in the field of chemiresistive, optical, and other sensors.

In the chapter presented by B. Lucas-Granados, R. Sánchez-Tovar, R. M. Fernández–Domene, and J. García-Antón we can see research results on novel nanostructured metal oxide photoanodes synthesized by electrochemical anodization of transition metals under hydrodynamic conditions. In electrochemical water splitting, upon illumination, the electric field generated by the surface band bending of a wide-bandgap n-type semiconductor drives the minority charge carriers toward the surface to oxidize water and produce electricity. Photoelectrochemical properties of the metal oxide anodes are discussed pertaining to the nanostructure peculiarities.

Catalysis is ubiquitous in almost any industrial chemical process and can be utilized to aid in the reduction of air pollution and generation of energy from hydrogen. The catalysts currently employed for environmental control are either noble metals or metal oxides, which are less expensive and readily processable. In the next chapter, C. Doroftei describes a series of nanostructured transition metal oxide compounds possessing perovskite structure that is applicable for catalytic combustion of volatile organic compounds at low and moderate temperatures. The synthesis method of the perovskite metal oxides is the precursor method of self-combustion with polyvinyl alcohol as a colloidal medium. The author discusses structural properties, morphology, and chemical composition of as-prepared perovskites in relation to the catalytic activity.

Authors R. Maric and H. Yu review in their chapter future perspectives of hydrogen as an energy carrier, classify water electrolysis technologies, and outline challenges in the development of membrane and catalyst materials. The oxygen evolution reaction at the anode is strongly influenced by the chemical properties and surface structure of the nanostructured catalyst. The study reveals that the co-presence of iridium oxides comprising Ir in III and IV oxidation states supports a higher oxygen evolution reaction activity due to the presence of electrophilic oxygen species, which promote the formation of O–O bonds. The size effects of bimetallic oxides and nanoparticles on the electronic structure of the Ir-based catalysts and the role of metal support to enhance oxygen evolution reaction activity and the catalyst stability are outlined.

Light emission and manipulation at the nanoscale is a long-standing research interest of the interdisciplinary scientific community. Although silicon dominates in present-day electronics, it is a poor emitter of light due to the indirect bandgap.

The following chapter presented by authors T. Takeuchi and Y. Horikoshi shows research results on the fabrication of amorphous Si/SiO_2 quantum wells, which incorporate silicon light-emitting nanocrystals and reports on the optical properties of these nanostructures. The light-emitting Si/SiO_2 nanostructures are nanolayers of amorphous Si, SiO_2, and Si/SiO_2 films grown by means of ultrahigh vacuum radio-frequency magnetron sputtering. The tentative absorption coefficients are evaluated for different thicknesses of silicon layers in Si/SiO_2 nanostructures. Quantum confinement effects are found to modify optical properties of the Si/SiO_2 nanostructures as inferred from the optical transmittance and reflectance spectra. The rapid-thermal-annealed Si/SiO_2 multilayer films demonstrate strong photoluminescence at ~1.45 eV originating from nanocrystalline silicon. The temperature dependence of the nanocrystalline luminescence intensity is discussed invoking the Kapoor model.

The last chapter by T. I. Shabatina reviews methods of supramolecular chemistry and cryochemistry applied to the synthesis and morphological and structural investigation of hybrid nanostructures and nanosized aggregates comprising plasmonic metals and several mesogenic compounds, which can reveal at certain temperatures different liquid crystalline mesophases—nematics, smectics, and cholesterics. The chapter also considers the directed formation of metal nanoparticles by applying thermal treatments. New hybrid nanosystems were characterised by infrared and UV-Vis spectroscopy, transmission electron microscopy, and selected area electron diffraction.

Finally, I wish to express my sincere gratitude to Ms. Romina Skomersic for support in the course of preparing this volume.

Yanina Fedorenko
The University of Liverpool, UK

Chapter 1

Introductory Chapter: Electron Microscopy - Research Highlights

Yanina Fedorenko

Additional information is available at the end of the chapter

http://dx.doi.org/10.5772/intechopen.80729

1. Introduction

Scientific knowledge begins with observation. Scientists use observation to obtain data and verify theories and hypotheses. The most straightforward method of observation consists in imaging an object under study. Therefore, microscopy is an integral part of modern research in natural sciences. Here, I highlight recent innovations of transmission electron microscopy such as the development of new techniques of sample preparation, application of electron diffraction to study atomic displacements in disordered materials, determination of local structural variations in glassy materials, and imaging dopants atoms in quantum dots. The highlights may interest researchers working on nanoscale phenomena in solid state physics, technology, materials physics, and engineering.

2. A novel method of sample preparation for transmission electron microscopy

Improved instrumentation and optics significantly contributed to the development of transmission electron microscopy (TEM) with considerable effort put into innovative techniques for preparation of samples and introduction of new equipment, which progressed from the ion-beam polishers to tripod polishing and focused-ion-beam systems. Even though the latter can damage a specimen, the method is almost universally used for the sample preparation in electron microscopy. To prevent sputtering damage, a photoresist or a combination of a photoresist and an inorganic film may be used.

The sample geometry for TEM analysis in a majority of studies is the thin-foil. A new technique of sample preparation has been developed to enable transmission microscopy in most basic equipment with the use of usual imaging modes [1]. As an example of TEM by specimen design, the strain in thin film planar devices has been studied. The strain is one of the important parameters for understanding nanoscale phenomena in physics because it controls the mechanical and electronic properties of materials. The new measurement technique of electron microscopy proposes an original design of the specimen geometry to apply the technique in basic conventional TEM, *i.e.* by exploiting the moiré imaging phenomenon. The Moiré patterns are analyzed for the strain, the lattice parameters, orientation, and distortions. A new method of two-sample preparation is proposed to utilize the properties of moiré patterns and improve the spatial resolution of strain maps. The method explores cutting a thin lamella sample into two pieces to provide precise control of parallelism of the two superimposed surfaces. The misorientation of the two surfaces is only 0.05° that is smaller by one order of magnitude than typical Bragg angles in the range of 0.3–0.6° and the zone, which needs to be imaged, is orientated almost in the same direction as the reference lattice. The spatial resolution of TEM imaging obtained by applying this sample preparation procedure is about one nanometer that is achieved by controlling rotations of the samples during preparation.

3. Disordered materials studied by electron transmission microscopy

The atoms vibrate in the matter at room temperature. The effect of vibrations is observable from diffraction patterns of electron microscopy. Vibrating atoms introduce a diffuse background of low intensity superimposed with the diffraction peaks. Without the thermal diffuse scattering (TDS) diffraction angles of stationary atoms would give a set of discrete lines. Electrons, as well as x-rays and neutrons, are characterized by a wavelength, which is dependent on an accelerating voltage applied between the cathode and the anode. A typical wavelength of an electron in a TEM instrument is 0.04–0.02 Å, *i.e.* by a few orders of magnitude smaller than the wavelength of x-rays. Therefore, much smaller Bragg angles can be achieved that enables crystalline order to be studied with higher precision. Also, x-ray analysis requires rotation of a sample, whereas electron diffraction does not. A range of disordered lattices can be studied by using electron transmission microscopy – flexible framework structures, disordered solid solutions, stacking fault structures, and materials susceptible to structural instabilities such as novel perovskite metal-halide semiconductors.

Diffuse scattering has been studied in the molecular crystals, which experience large molecular displacements as a result of thermal molecular vibrations [2]. The dynamic disorder in molecular systems contributes to fluctuations of the transfer integrals, *i.e.* to charge carrier transport in disordered molecular solids and the energy distribution of transport sites. The lattice vibrational modes of a range of organic semiconductors have been studied from electron diffraction patterns in TEM. The research illuminates the interrelation of thermal lattice fluctuations and molecular structure of organic semiconductors.

4. Spatial heterogeneity in metallic glasses studied by high-angle annular dark field microscopy

Glasses have been considered as liquids frozen by quenching of high-temperature melts. Metal glasses are hard to prepare because compared to organic materials and silicates metals readily crystallize during solidification that introduces heterogeneities in the amorphous materials structure. The heterogeneous nature of metal glasses has long escaped an experimental evidence since nanoscale crystal-like aggregates in amorphous solids appear at a very short length scale, which is not detectable by high-resolution electron diffraction methods.

To reveal crystal-like aggregates at nano-scale and study the local atomic structure of spatial heterogeneous metallic glasses, researchers use advanced experimental methods of angstrom-beam electron diffraction (ABED) and aberration-corrected scanning transmission electron microscopy (STEM) [3]. The crystalline phase is detected by the technique of high-angle annular dark field (HAADF) imaging in STEM. In HAADF, the scattering angles are large and incoherent thermal diffuse scattering surpasses the coherent Bragg scattering. Being sensitive to the atomic numbers, HAADF provides contrast images of compositional non-uniformities in materials. The structurally different regions in the metallic glass are observed as dark and bright areas in HAADF STEM images and analyzed by using ABED to detect the local atomic configurations. The structural variations within the glassy material result in spatial heterogeneity of metallic glasses that manifest itself in a fluctuation of density occurring without fluctuation of chemical composition. The structural heterogeneity has important implications in glass science and technology because many of the material properties such as the glass-forming ability and the elastic behavior are structurally dependent.

5. Ultra-thin graphene oxide support film for TEM observations of core-shell nanocrystals

Over the last five decades, SiO_2 has been intensively studied as the basic ingredient in field-effect transistors, silicon photovoltaics, and nanoscale light-emitting sources. Metal-oxide-semiconductor technology is dominating for very large scale integration applications, enabling smaller dimensions, a faster switching speed, and a lower power dissipation. Ability to control the oxide thickness with an angstrom precision facilitates the fabrication of ultra-small devices operating at a higher drive current. Despite the successful implementation of planar SiO_2/Si nanostructures in transistors, the device application of Si nanocrystals in electronic and optoelectronic devices is still hindered due to low carrier concentration and low charge carrier mobility. Although the optical and electronic properties of silicon nanostructures are remarkably dependent on the modification of the defect ensemble at the Si nanocrystal-SiO_2 interface, direct imaging of the amorphous structure surrounding silicon nanocrystals has been found difficult due to the insufficient contrast of electron microscopy. The use of atomically thin graphene oxide supports in electron microscopy of colloidal silicon quantum dots reveals the amorphous structure of the surrounding core with the core thickness being affected by the concentration of doping atoms [4]. The insufficient doping might stem from the aggregation of the dopant and the silicon atoms in a partially oxidized amorphous shell, which forms on the top of a crystalline silicon core.

Author details

Yanina Fedorenko

Address all correspondence to: janina.fedorenko@gmail.com

Stephenson Institute for Renewable Energy, University of Liverpool, Liverpool, UK

References

[1] Cherkashin N, Denneulin T, Hÿtch MJ. Electron microscopy by specimen design: Application to strain measurements. Scientific Reports. 2017;**7**:12394. DOI: 10.1038/s41598-017-12695-8

[2] Eggeman AS, Illig S, Troisi A, Sirringhaus H, Midgley PA. Direct measurement of molecular motion in organic semiconductors. Nature Materials. 2013;**12**:1045-1049. DOI: 10.1038/nmat3710

[3] Zhu F, Hirata A, Liu P, Song S, Tian Y, Han J, Fujita T, Chen M. Correlation between local structure order and spatial heterogeneity in a metallic glass. Physical Review Letters. 2017;**119**:215501. DOI: 10.1103/PhysRevLett.119.215501

[4] Sugimoto H, Yamamura M, Sakiyama M, Fujiia M. Visualizing a core–shell structure of heavily doped silicon quantum dots by electron microscopy using an atomically thin support film. Nanoscale. 2018;**10**:7357-7362. DOI: 10.1039/C7NR09474D

Chapter 2

Molecular Simulation Study of Water Transport through Aquaporin-Inspired Pore Geometry

Daejoong Kim and Majid Shahbabaei

Additional information is available at the end of the chapter

http://dx.doi.org/10.5772/intechopen.77394

Abstract

As the solid-state nanochannels are hourglass-shaped as opposed to having a straight structure, little effort has been made on the transport phenomena across aquaporin-mimicking ultra-fine pores. Moreover, to the best of the authors' knowledge, the existing studies on the water transportation through hourglass-shaped pore structures are limited to the symmetric geometries. In order to benefit from the advantage of combining the properties of nanoporous graphene and the unique structure of the aquaporin water channels, this chapter aims to study designing a highly effective permeable membrane for water transportation through nanopores with hourglass shape in multilayer structure using molecular dynamics (MD) simulation. Keeping the layer separation constant at 6 Å, we intend to examine the geometry effect on the pressure-induced transport of water molecules through hourglass-shaped nanopores. Three models of hourglass-shaped nanopores are selected: a symmetric structure $\alpha = 0$ and asymmetric structure of $\alpha = 1/3$ and $\alpha = -1/3$. Our MD findings indicate that the permeability of water molecules highly increases across hourglass-shaped pores with an asymmetric structure of $\alpha = 1/3$ because of the increased number of hydrogen bonds resulting from the length effect. Thus, we can conclude that an hourglass-shaped pore with a proper design can remarkably improve the rate of water transportation.

Keywords: water transport, aquaporin water channel, hourglass-shaped pore, molecular dynamics simulation

1. Introduction

The growing demand for water in everyday human activities makes producing fresh water a great challenge. Using membrane separation technology in the areas of water treatment is widely implemented. Producing portable water from existing resources such as surface water, seawater, brackish water, or groundwater can be done by using membrane processes. The membrane acts as a barrier with which two phases are separated, and transport of different chemical species are restricted selectively. It is basically referred to as a thin layer which separates materials depending on their chemical/physical properties when an external force (e.g., gradient pressure, electrical force, etc.) is applied across the membrane. Nanoporous graphene has gained great attention as promising membrane material for RO desalination because of its exceptional properties such as being mechanically strong, chemically robust, and extremely thin. Furthermore, using graphene—with such an extremely thin thickness—in desalination plants can remarkably yield higher water fluxes with significant reduction in the operating costs, for example, cost in terms of energy and cost in terms of capital and infrastructure requirements [1]. Considering the graphene-based membrane as highly selective and permeable, the generation of pore with various pore shape and size can be achieved in graphene by using electron beam irradiation [2]. Recent studies in both theory and experiment revealed that graphene [3–8] and graphene oxide [9, 10] were highly selective and permeable and could be used as filtration membranes. Though single-layer graphene has demonstrated to be a promising candidate as a reverse osmosis membrane, synthesizing multilayer graphene is economically affordable [11–14]. Considering parameters including pore size and chemistry, a wide range of studies have considered transport phenomena through single-layer graphene. However, less effort has been made on transport properties through nanoporous multilayer graphene. Experimentally, the studies on graphene oxide have revealed that constructing carbon-based material in molecular-scale porous structure could be achieved with tuned pore sizes and interlayer spacing [15–18]. A better selectivity and higher permeability for artificial membranes could be achieved mimicking bio-inspired membranes such as aquaporins. Aquaporins are pore-forming proteins and ubiquitous in living cells, which provide fast water transportation in a selective manner resulting from their unique structure of hourglass shape [19]. Recent study [20] showed that the desalination performance could be significantly improved incorporating the properties of aquaporin water channel into desalination membranes. Revealing an hourglass-shaped structure for the solid-state pores [21], transport phenomena pores in hourglass-shaped structure is not fully understood. Although the existing studies on water transportation properties across hourglass-shaped pore systems are limited to the symmetrical structures, to the best of our knowledge [22–24], further molecular-level studies on the transport phenomena across asymmetrical geometries are still needed [25–27]. The water transportation phenomena in confined environments (e.g., nanochannels), where the critical dimension can be a few molecular diameters, differ from macroscopic scales which explains the transport phenomena based on continuum theories (e.g., Navier-Stokes equations). Accordingly, a much-needed molecular-level understanding of the water transportation phenomena at the nanoscale is necessary. Molecular dynamics (MD) has become a powerful tool for investigating the effects of confinement on fluid properties and calculating phenomenological transport coefficients for use in multi-scale fluid flow models. A significant effort has been devoted, often with the help of MD simulation, to understand the water transportation properties in carbon nanotubes (CNTs) and graphene [28–31]. Several researchers have shown a freezing state for water confined in CNTs at low temperature and the appearance of ice phases stabilizing by confinement [32, 33]. Using MD simulations, Koga and colleagues [32] showed that water molecules inside CNTs with diameters varying from 11 to 14 Å indicate phase transitions with applied pressure of 50–550 MPa. Experimentally, Kolesnikov and colleagues [33] observed a combination of water-chain and ice-shell structure for water molecules confined in CNTs having a liquid phase state even at low temperature. The appearance of single-layer and bi-layer ice structures were also disclosed for water molecules confining in between graphene slits under applied pressure [34, 35]. Another study revealed that water confined in graphene slits freezes at ambient temperature while no pressure is applied [36]. Chakraborty and colleagues [37] have shown a diversity of results on the structure, dynamics, and thermodynamics of confined water molecules in CNTs and graphene oxide sheets. They found the appearance of the single-file ordered structure for water molecules confined in narrow CNTs and the appearance of the n-gonal ice structure in larger CNTs. Their findings also revealed the appearance of two-dimensional ordered structures for water molecules confined in slit-pore geometries. Considering the graphene properties combining with aquaporin water channel structure, this study aims to design an optimal and effective permeable membrane for water transportation across hourglass-shaped pores in multilayer nanoporous graphene using MD technique.

2. Simulation setup

Molecular dynamics (MD) is used to investigate the transport of water molecules across hourglass-shaped nanopores in nanoporous multilayer graphene. The schematic of the simulation models are given in **Figure 1**. Three pore models of hourglass shape of asymmetric (α = 1/3 and α = −1/3) and symmetric (α = 0) are chosen. Seven layers of graphene (31.9 × 34.3 Å^2) with a constant interlayer separation of d = 6 Å with total length of 36 Å is represented as the multilayer structure. Considering the single-file water arrangement across the narrowest part and a bulk structure of water molecules in the largest segments, we considered being the narrowest (a) and the largest diameters (D) of hourglass-shaped structure in 4 and 7 Å, respectively, embedding in a periodic box of over 2000 SPC/E water molecules. To generate the pore in graphene layers, we removed adjacent carbon atoms in the center and considered the pores with no sharp edges (for more information, see Ref. [12]). To generate water flow in the models, we applied the pressure difference along −z direction across α = 1/3, α = −1/3, and α = 0 systems. A pressure-induced flow, ΔP = nf/A (denoting n and A as the number of water molecules and the membrane surface area, respectively), is achieved by applying a constant force, f, on water molecules inside the reservoir [38]. All the simulations are performed using large-scale atomic/molecular massively parallel simulator (LAMMPS) simulation package [39], implementing the constant number (N), volume (V), and energy (E) (NVE) ensemble with dissipative particle dynamic (DPD) thermostat [40] aiming at maintaining the temperature constant at 300 K. We have chosen a time step of 0.5 fs in order for ensuring that water molecules between the bulk region and layers are accurately simulated. Considering a periodic boundary condition in three directions, a cutoff radius of 10 Å is selected for the Lennard-Jones (LJ) potentials. To reduce the computational cost, we have considered the carbon atoms of the graphene to be frozen. We have modeled only the interaction between water molecules (oxygen) and the graphene (C) by using the Lennard-Jones parameters as σ_{O-O} = 3.169 Å, ε_{O-O} = 0.1515 kcal/mol, σ_{O-C} = 3.190 Å, and ε_{O-C} = 0.0956 kcal/mol. Using the Ewald summation method [41] for the coulomb interactions, the SHAKE algorithm is applied to hold the water molecules rigid [42]. Conducting more than 15 ns, MD runs for every three cases (α = 1/3, α = −1/3, and α = 0), the last 12 ns trajectory was considered for data analysis.

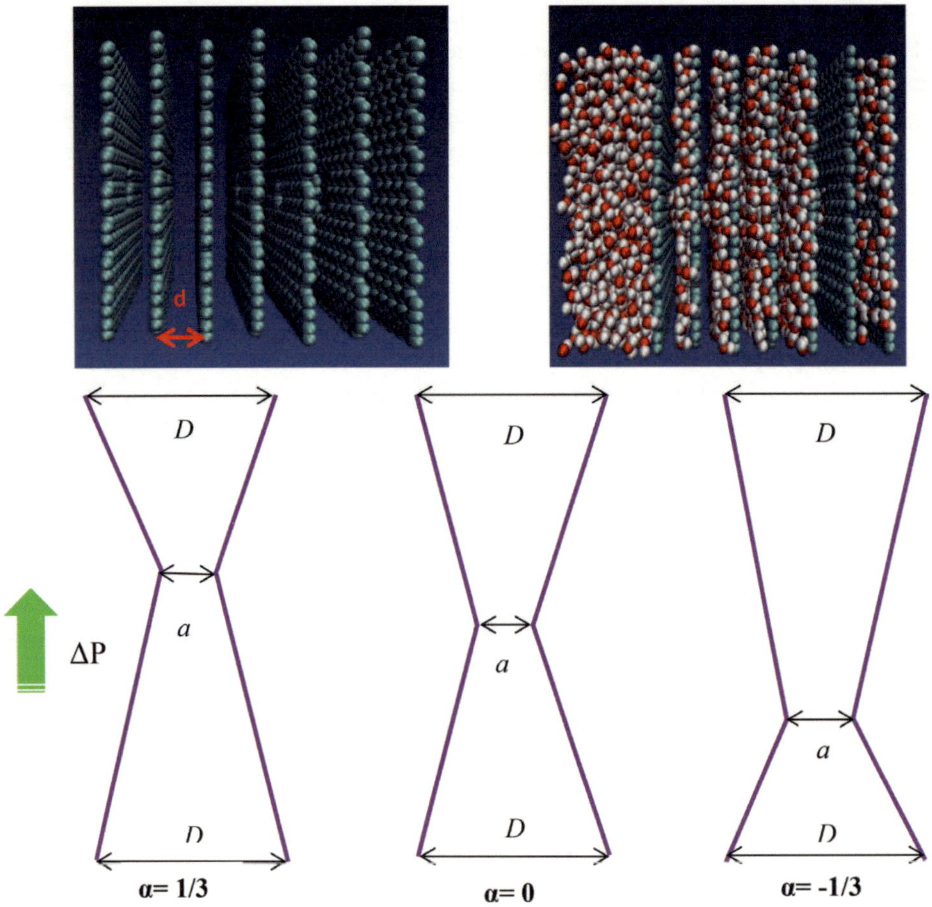

Figure 1. Schematic diagram of the asymmetric ($\alpha = 1/3$ and $\alpha = -1/3$) and symmetric ($\alpha = 0$) hourglass-shaped nanopore in multilayer structure with a constant interlayer spacing of d = 6 Å. The hourglass-shaped pores with a total length of 36 Å, the largest diameter D = 7 Å, and the narrowest diameter a = 4 Å are considered as the pore models. A pressure difference is applied along −z direction to drive water flow.

3. Results and discussion

In our simulations, we applied the pressure difference, ΔP, ranging from 365 to 635 MPa across the multilayer nanopore systems. Three pore models of $\alpha = 1/3$, $\alpha = -1/3$, and $\alpha = 0$ with the long, short, and equal entrance lengths are considered. **Figure 2a** shows the average number of water occupancy with the applied pressures inside each of the three models. The water occupancy is calculated by counting the number of water molecules, which could diffuse through the 36-Å-long multilayer systems during the simulations. It has been found that the average water occupancy in the case of $\alpha = 1/3$ is higher than that of $\alpha = -1/3$, and $\alpha = 0$ systems. Since the pore volume increases, at $\alpha = 1/3$, for example, the influx of water molecules could be encouraged stemming from an increased cone length. Though a significant variation of the number of water occupancy in relevance to applied pressure is found in $\alpha = -1/3$ model, generally, the

http://dx.doi.org/10.5772/intechopen.77394

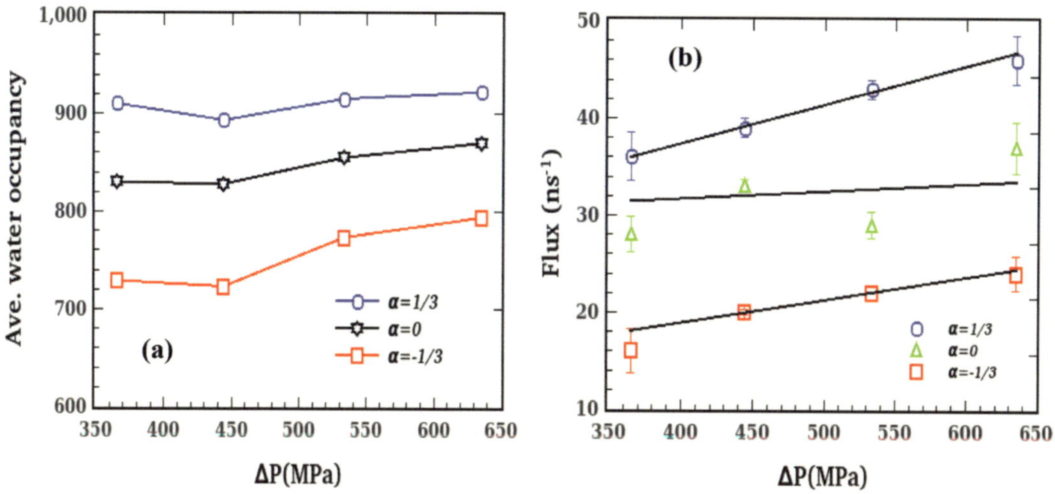

Figure 2. (a) Average water occupancy inside the pore systems and (b) value of water flux through the pore models shown as a function of the pressure difference.

value of water occupancy inside the pores is more affected by the pore structure compared to the applied pressure (e.g., $\alpha = 1/3 > \alpha = 0 > \alpha = -1/3$). The number of water flux is calculated by counting the number of water molecules that cross the pore thoroughly, as shown in **Figure 2b**. Qualitatively, the results indicate that the water flux increases linearly with applied pressure, which is consistent with the recent study [4]. Moreover, the MD results show that the number of water flux is higher in the case of $\alpha = 1/3$ compared to $\alpha = 0$ and $\alpha = -1/3$ systems. Interestingly, the symmetric hourglass-shaped pore ($\alpha = 0$) shows a lower flux of water molecule compared to the asymmetric pore structure (e.g., $\alpha = 1/3$), which attributes to the entrance effects. Therefore, it is understood that the entrance effects have a significant contribution to flux variation inside pores with the structure of hourglass shapes. The permeability of water molecules through pores in hourglass shape is significantly restricted due to the energy barriers at the entrances. In case of $\alpha = 1/3$ for example, a large amount of water molecules entering the long cone can contribute to reduce the energy barriers caused by the narrowest section of pore (of which accommodates water molecules only in single-file configuration), resulting from the increase in the number of hydrogen bonds. It is also inferred that the decrease in the cone angle would enhance the permeability of water molecules across hourglass-shaped pores in a multilayer structure (indeed, this implication very recently discussed for the permeability of water molecules in solid-state nanochannels of hourglass shape [22]). To assess the transport capacity of a membrane, one could calculate an important parameter named osmotic permeability. By drawing a best-fit slope through flux versus applied pressures, the osmotic permeability of a membrane can be achieved as shown in **Figure 2b**. In agreement with the results reported for flux, the corresponding magnitudes are $(4 \pm 1) \times 10^{-14}$, $(7.52 \pm 9.76) \times 10^{-15}$, and (2.3 $37 \pm 4.33) \times 10^{-15}$ $cm^3\ Pa^{-1}\ s^{-1}$ for the pore systems of $\alpha = 1/3$, $\alpha = 0$, and $\alpha = -1/3$, respectively. It is also inferred that achieving high permeability of water molecules across the pore of hourglass shape would strongly depend on the structure of the pore. Therefore, a design of hourglass-shaped pore with which adsorbs bulk water molecules can be a promising geometry aimed

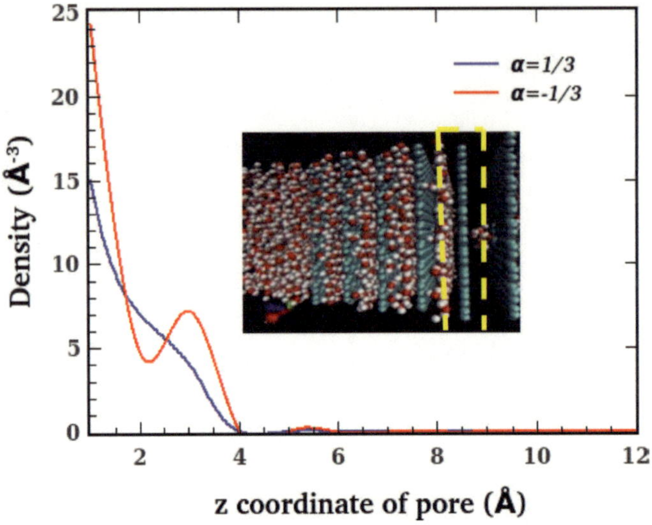

Figure 3. Density of water molecules along the axial direction of the layer with the narrowest diameter of the pore inside $\alpha = 1/3$ and $\alpha = -1/3$. The inset shows the layer with the narrowest pore (dashed line).

at obtaining high permeability. Within this context, the water molecules in bulk-like manner contribute to remarkably reduce the energy barriers inside the pores because of the size effects. Next, we investigate the viscosity dissipation inside asymmetric pore systems at the smallest section of the pores. Very recently, a numerical study [22] has shown that the entrance effects (defined as viscous dissipation) would dramatically restrict the water permeability inside solid-state pores of hourglass shape. Basically, the molecules of water confined in graphene nanopores show different viscosity compared to the carbon nanotubes (CNTs). Considering a similar characteristic of pore size, a recent study has shown that the molecules of water confining in a graphene pore pose higher viscosity compared to the water molecules confining in CNTs. Increasing the pore diameter leads to the increase in the viscosity of confined water molecules in CNTs approaching the bulk viscosity [29, 43]. In graphene nanopores, the viscosity of water molecules increases, while decreasing the pore diameter in accordance with the structure of water molecules [38]. The structure of water molecules along the flow direction inside graphene nanopores shows different behavior compared to CNTs' nanochannels. Recent MD study [44] has shown that the viscosity of water molecules across graphene nanopore is linearly increased in accordance to the layering effect. In fact, the viscosity of water molecules through graphene nanopores is in relation to the layering effect of water molecules. Through the density distribution, one can obtain the water layering effect along the flow direction in the graphene system. Here, we measured the water layering factor as $(\varrho_{max} - \varrho_{min})/\varrho_{bulk}$, to quantify the water layering effect inside only the asymmetric pore structures, noting that ϱmax and ϱ_{min} are the maximum and minimum density numbers in the vicinity of the layer with the narrowest pore diameter, respectively, as shown in **Figure 3**. The ϱ_{bulk} is estimated to be in the range of bulk fluid densities reported, $0.8\sigma_{O-O}^{-3} \leq \varrho_{bulk} \leq 1.1\sigma_{O-O}^{-3}$ [45]. When $\alpha = -1/3$, the density profile near the layer with the smallest pore diameter shows greater distribution compared to $\alpha = 1/3$, suggesting a significant effect of water layering inside $\alpha = -1/3$. Within

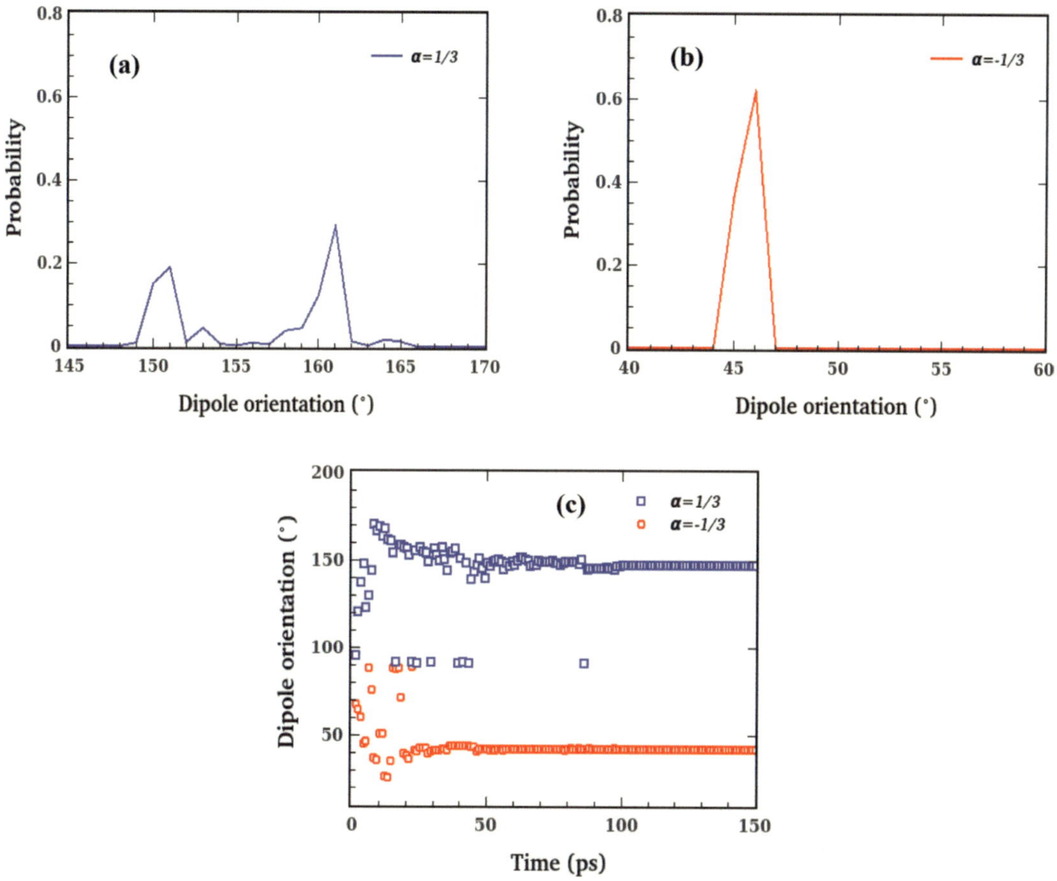

Figure 4. Probability distribution of dipole orientation of water molecules inside the (a) $\alpha = 1/3$ and (b) $\alpha = -1/3$ systems, and (c) dipole orientation of water molecules in relation to time inside these two pore systems.

this context, the water layering effect increases the viscosity of water molecules while crossing the smallest diameter in single-file arrangement. Accordingly, a significant reduction in the local permeability inside $\alpha = -1/3$ is expected.

Furthermore, in case of $\alpha = 1/3$, large amount of water molecules (bulk-like) approaching the narrowest diameter of the pore should contribute to diminish highly the water viscosity of single-file configuration because of the increase in hydrogen bonds. Although the dipole orientation of water molecules inside CNTs maintains oversimulation time, in graphene nanopores, the dipole orientation of water molecules is frequently flipped [46]. Here, we intend to show how dipole orientation of water molecules may behave across pores in multilayer system. Taking into account all the water molecules inside pores, the averaged dipole orientation was estimated by computing the angle between water dipole and the axial direction (perpendicular to the surface of the multilayer). **Figure 4a** and **b** shows the profiles of the averaged dipole orientation of water molecules inside asymmetric pores $\alpha = 1/3$ and $\alpha = -1/3$, respectively. Over 4000 calculated values are used to plot the average dipole orientation of water molecules inside asymmetric structures ($\alpha = 1/3$ and $\alpha = -1/3$). In case of $\alpha = 1/3$, the average dipole orientation fell into two range groups of 149–153° and 158–162°, while in the case of $\alpha = -1/3$, the value of average dipole orientation fell only in the range of 44–47°. This implies a degree of water molecule ordering inside the pore in the case of $\alpha = 1/3$, which enhances the formation of hydrogen bonds between water molecules. Moreover, the variation of the dipole orientation tends to strengthen the water-water interaction resulting in higher hydrogen bonds. This is responsible for fast transport of water molecules in monolayer graphene [13]. We believe that this implication could be also valid in multilayer graphene systems.

Although, the average dipole orientation inside $\alpha = 1/3$ and $\alpha = -1/3$ systems is only flipped during the course of few picoseconds, it maintains for whole the simulation time showing a higher magnitude of dipole inside $\alpha = 1/3$, as given in **Figure 4c**. At constant layer spacing (d = 6 Å), the obtained results show that the dipole orientation of water molecules inside the multilayer graphene membrane is different from that across single-layer graphene, in which the dipole orientation has frequently flipped. This implies that every single layer in multilayer system provides the energy barrier which allows the water molecules to freely flip. It would be expected that by increasing the layer spacing up to several angstroms, this phenomenon changes. The fluctuation of free energy occupancy in relation to the number of water molecules across asymmetric pores is given in **Figure 5a** and **b**. We computed the free energy using the equation, $G(N) = -Ln[P(N)]/b$, where P(N) is the probability of finding the exact N molecules of water in the system, $b = (k_BT)^{-1}$, K_B and T stand for the Boltzmann constant and temperature, respectively. Ranging from 896 to 926 water molecules, the most probable of water molecules is being with 904 molecules inside $\alpha = 1/3$. However, considering 722–756 water molecules inside $\alpha = -1/3$, the 728 is the most probable. In comparison to $\alpha = -1/3$, the water molecules are increasingly accommodated inside $\alpha = 1/3$ because of the strong interplay between molecules, which results in forming the stabilized chain of water molecules. In comparison to $\alpha = -1/3$, a wider distribution of probability with higher number of water molecules are found inside $\alpha = 1/3$. As a result, a high rate of water conductivity can be found inside fully filled asymmetric model of $\alpha = 1/3$. In addition, the probability of existence of higher numbers of water molecules is scarcely observed inside $\alpha = -1/3$ due to the strong water-wall interplay.

The free energy of occupancy in relation to the hydrogen bonds number is given in **Figure 5c**. The hydrogen bond was calculated in accordance to the method reported in Ref. [47]. As the figure shows, the average number of hydrogen bonds varies from 1.48 to 1.61 inside $\alpha = -1/3$, whereas it ranges from 1.56 to 1.73 for the case of $\alpha = 1/3$. Then, a wider distribution for probability of formation of higher numbers of hydrogen bonds inside $\alpha = 1/3$ suggests an enhancement in the water density distribution. In other word, the establishment of the stable chains of water molecules inside $\alpha = 1/3$ can support formation of the long-lasting network of hydrogen bonds which is significantly contributed to achieving higher flux. Radial distribution function (RDF) is an important parameter to analyze the structure of molecules in confined environments. **Figure 6a** depicts the profile of the RDFs between the water molecules (oxygen atoms) inside asymmetric pore models. A rapid decay for the RDFs can be seen inside both pore models of $\alpha = 1/3$ and $\alpha = -1/3$. According to the high magnitudes of the RDFs at low interatomic distances, the formation of solid clustering between water molecules inside $\alpha = 1/3$ is predicted, which triggers potential reduction in the energy barrier caused by the vacant spaces between molecules, as shown in **Figure 6b**. Furthermore, these results of the RDF are in agreement with the results of the free energy fluctuations. The transport mechanism of molecules in confined spaces could be monitored by using the potential mean force (PMF). Using the RDF which illustrates how density varies as a function of distance from a reference particle, one can calculate the PMF since the density would be a key index to explain probability [48]. **Figure 6b** illustrates the profiles of the potential mean forces (PMFs) for water molecules inside $\alpha = 1/3$ and $\alpha = -1/3$ pore systems. Because there is a weak interaction between the molecules of water inside $\alpha = -1/3$ pore model, a large energy barrier caused by vacant spaces between the molecules of water is understood. We believe that the large energy barrier could be attributed to the increased viscosity caused by the pore size at the smallest section, which dramatically reduces the passage of molecules of water. Encouraging the water molecules in bulk structure to get through the pore of $\alpha = 1/3$, they can significantly contribute overcoming the energy barrier. Therefore, water molecules with a tight hydrogen bonding network can pass the smallest section of the pore, which contributes strongly improving the entire passage of molecules of water across the pore. To further understand the mechanism of water transportation in confined

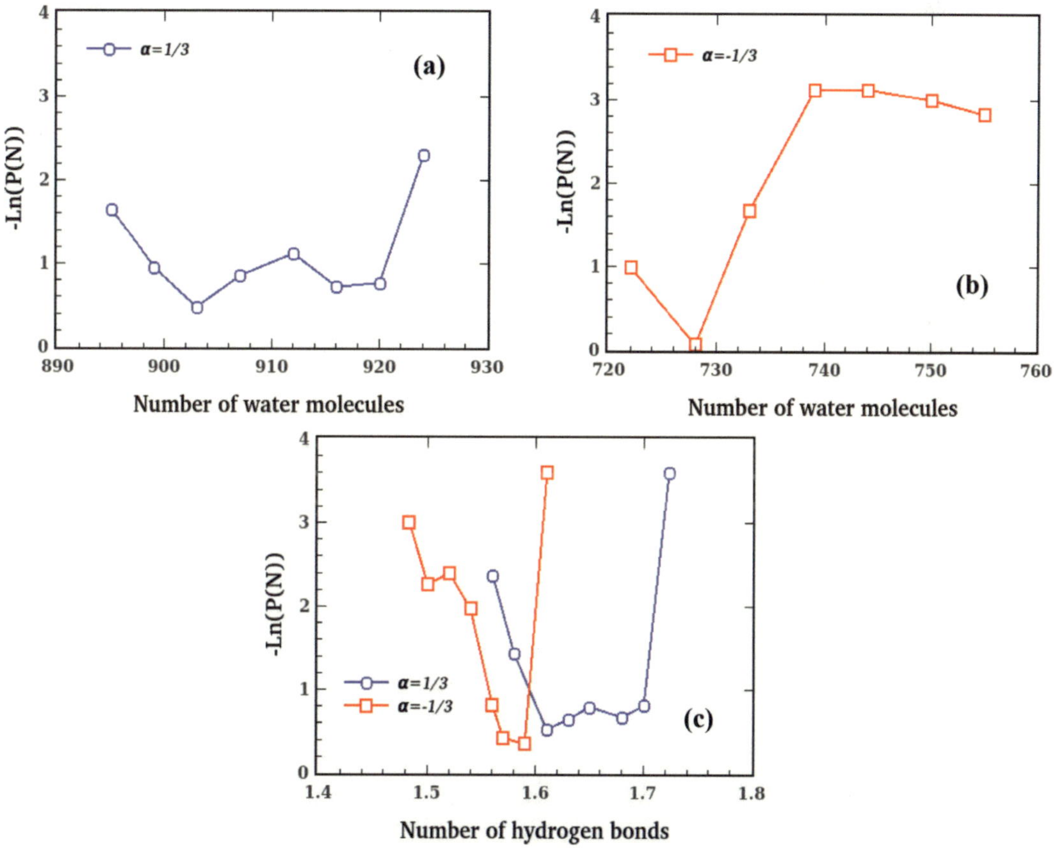

Figure 5. Free energy of occupancy as a function of the number of water molecules inside the (a) $\alpha = 1/3$ and (b) $\alpha = -1/3$ systems, and (c) free energy of occupancy as a function of the number of hydrogen bonds inside these two pore models.

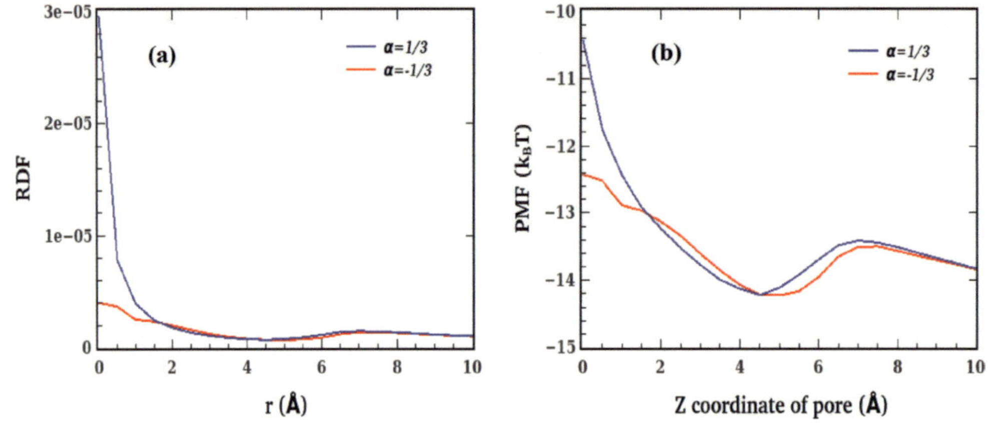

Figure 6. (a) Profiles of RDF between the oxygen of water molecules inside whole systems of $\alpha = 1/3$ and $\alpha = -1/3$ pore models and (b) profiles of the potential mean force (PMF) for water molecules across the whole systems of $\alpha = 1/3$ and $\alpha = -1/3$.

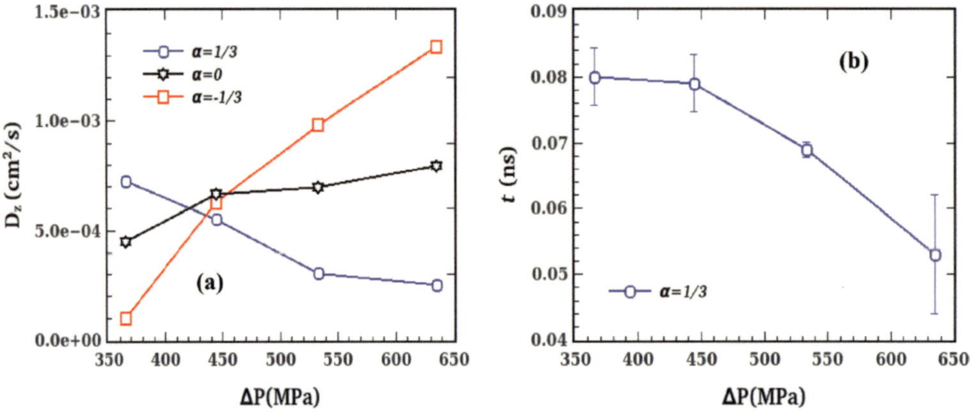

Figure 7. (a) Axial diffusion coefficient of water molecules inside symmetric ($\alpha = 0$) and asymmetric ($\alpha = 1/3$ and $\alpha = -1/3$) pore systems and (b) average residence time for water molecules inside $\alpha = 1/3$, as a function of the pressure difference.

geometries, an important parameter known as diffusion coefficient can be used. We computed the axial diffusion coefficient by calculating the mean square displacement (MSD), as given in Ref. [49]. The diffusion rate is dependent on density because it stems from the collision between molecules and the interplay between molecules and the pore surface [50]. The axial diffusion coefficient of water molecules in the −z coordinate of the symmetric and asymmetric pore systems is given in **Figure 7a**. The figure shows that the axial diffusion coefficient decreases with applied pressure inside $\alpha = 1/3$; however, it is increased inside $\alpha = -1/3$ and $\alpha = 0$ pore systems indicating a larger slope for diffusion inside $\alpha = -1/3$. It can be found that the increase in the density with applied pressure could be the responsible for the increase in the diffusion coefficient of water molecules inside $\alpha = -1/3$ and $\alpha = 0$ pore models. Looking at the low density states (in relation to low applied pressures), we believe that the water layering would be extremely small and the water molecules confined in between the layers would provide very limited spaces for molecules to move due to the strong interaction between water layer surfaces. Considering states with higher densities, the large number of hydrogen bonds would contribute to form the larger clusters yielding larger movement. In case of $\alpha = 1/3$ pore, by increasing the applied pressure, the diffusion coefficient decreases. We believe that the water density does not have a significant effect on the mechanism of diffusion across $\alpha = 1/3$ pore, as can be realized through the result reported for the water occupancy (see **Figure 2a**). The decreased diffusion coefficient with applied pressure could be ascribed to the formation of stabilized hydrogen bonding networks resulting from weak interaction between water layer surfaces. In comparison with $\alpha = 0$ and $\alpha = 1/3$ pore systems, a sharp slope for diffusion coefficient is observed inside $\alpha = -1/3$ which suggests a density-dependent effect. Since the applied pressure increases, a collective diffusion mechanism can be understood inside $\alpha = 1/3$ because of the formation of the stabilized hydrogen bonding networks. A discreet mechanism for filling behavior of water molecules with applied pressure can be realized across $\alpha = -1/3$ and $\alpha = 0$ systems due to existence of the large energy barrier. At the overlap pressure (444 MPa), the similar magnitudes for the diffusion coefficient of water molecules inside each of the three pore systems are obtained. The average residence time of water molecules inside $\alpha = 1/3$ system as a function of the pressure difference is given in **Figure 7b**. Considering a molecule entering from one side and leaving the pore from the other side, one can calculate the residence time $t_{res} = t_{out} - t_{in}$. In **Figure 7b**, we can see that the residence time inside $\alpha = 1/3$ shows a negligible variation when the applied pressure increases from 365 to 444 MPa, whereas it highly decreases while the applied pressure increases from 444 to 635 MPa which suggests the complete passage of a molecule across the pore in a short course of simulation time. Accordingly, it is inferred that the time residence is dependent on the applied pressure. Since the applied pressure increases, water molecules in bulk structure are encouraged to the cone and exit the pore in a fast behavior. This suggests that the structure of the pore has less contribution to the movement of water molecules. The profiles of the axial and total velocities of water molecules inside pores are shown in **Figure 8a** and **b** as the function of applied pressure. By increasing the applied pressure, the axial velocity of water molecules is strongly increased inside $\alpha = 1/3$, whereas a slight increase is found for the axial velocity inside $\alpha = 0$ system. In case of $\alpha = -1/3$, the increase in the applied pressure results in the reduction in the axial velocity inside the pore. It can be found that by increasing the applied pressure, the occupancy number of water molecules inside $\alpha = -1/3$ is increased. Due to the strong interaction between water layer surfaces it should be difficult for the large displacement of water molecules across the pore model of $\alpha = -1/3$, which results in producing low axial velocity. In addition, the figure describes high speed of transport of water molecules with applied pressure across $\alpha = 1/3$ and $\alpha = 0$ systems because of the weak interaction of water layer surfaces and the strong water–water interaction. Accordingly, the formation of tight hydrogen bonding networks is being expected inside $\alpha = 1/3$ and $\alpha = 0$ pores. It is realized that the pore structure plays a dramatic role on the transport phenomena across pores in hourglass structure. **Figure 8b** indicates the profile of the total velocities of water molecules in relation to applied pressure inside each of the three pore models. Generally, the total velocity (Vx + Vy + Vz) is higher inside $\alpha = -1/3$ compared to $\alpha = 1/3$ and $\alpha = 0$ systems. At applied pressure of 365 MPa, our simulation results show the same value for the total velocity inside symmetric and asymmetric pore systems. It is understood that at given applied pressure, the velocity in the x and y

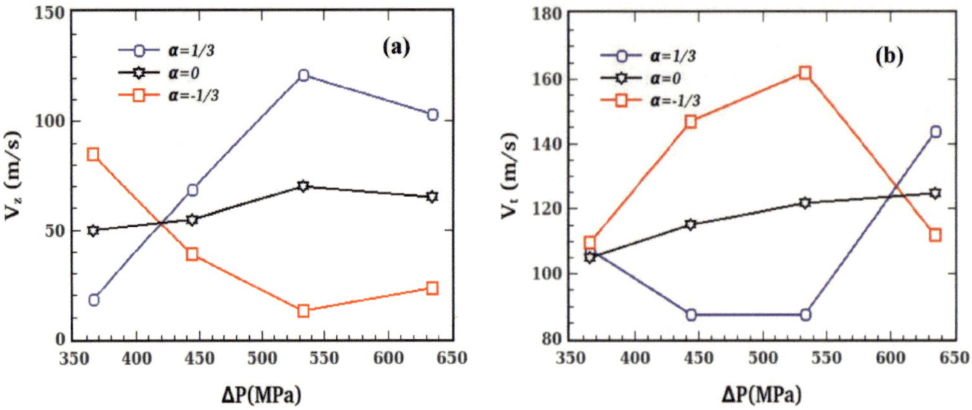

Figure 8. (a) Axial velocity and (b) total velocity of water molecules across $\alpha = 1/3$, $\alpha = 0$, and $\alpha = -1/3$ systems.

directions has a significant impact on total velocity inside $\alpha = -1/3$ model compared to $\alpha = 1/3$ and $\alpha = 0$ structures. The results reported in **Figure 8a** and **b** together suggest that the contribution of velocity in the z direction (Vz) should indicate the permeation ability in our simulation models, though a negligible effect of the velocities in x and y directions on the transport mechanism could be understood.

4. Concluding remarks

Graphene with one-atom thickness of sp^2 carbon atoms organizing a honeycomb structure possess an excellent mechanical strength, supreme electrical and thermal conductivity, and other amazing properties. Lately, the application of nanoporous graphene in desalination and water purification processes has been greatly increased. Using experimental techniques [51–54], one can generate nano-/micro-scale pores of various diameters and shapes on graphene sheets. Considering the properties of nanoporous graphene and taking advantage of aquaporin water channel structure, this computer simulation study aims to design an optimal and effective permeable membrane for water transportation across hourglass-shaped pores in multilayer nanoporous graphene.

In this study, by conducting MD simulations, we investigated the mechanism of water transportation across hourglass-shaped nanopores in symmetric and asymmetric structures constructed by multilayer nanoporous graphene. Considering a constant interlayer spacing of 6 Å, we particularly focused on the applied pressure effects on the transport properties through symmetric ($\alpha = 0$) and asymmetric ($\alpha = 1/3$ and $\alpha = -1/3$) pores with hourglass shape in multilayer graphene. The results show a higher occupancy of water molecules in relation to applied pressure inside $\alpha = 1/3$ compared to $\alpha = -0$ and $\alpha = -1/3$. In case of $\alpha = 1/3$, because of the increase in the volume of the pore (the increase in the cone length), the flow of molecules of water is fostered. Then, the occupancy of water molecules inside the pore with hourglass shape is geometry dependent compared to pressure effect. The number of flux of water increases linearly as a function of applied pressure inside the pores, suggesting higher flux inside $\alpha = 1/3$ pore. In comparison with $\alpha = -1/3$, the water molecules crossing the smallest pore diameter inside $\alpha = -1/3$ system have shown a larger layering effect which strongly reduce the water transport rate. The findings revealed that the asymmetric model of $\alpha = 1/3$ is able to transfer a larger number of fluxes in comparison to the symmetric pore of hourglass shape ($\alpha = 0$). This could be ascribed to the small viscosity in relation to small water layering which comes from the length effects. It is also understood a proportionality between the permeability and the pore structure with a higher value for the case of $\alpha = 1/3$. In comparison with $\alpha = -1/3$, the MD trajectory indicated that the dipole orientation is widely distributed inside $\alpha = 1/3$, falling in the ranges of 149–153° and 158–162°. Basically, a large distribution of dipole orientation triggers the fast flow of water inside the pore. It can be seen that the diffusion coefficient decreases with the pressure difference inside $\alpha = 1/3$ pore because of the strong clustering of water molecules. A collective and discreet diffusion mode is being inferred for transport mechanisms inside $\alpha = 1/3$ and $\alpha = -1/3$ systems, respectively. Finally, our MD results described that the contribution of the velocity perpendicular to the layers (z-axis) exposes the influx ability of molecules of water inside the pores.

Acknowledgements

This work was supported by the National Research Foundation of Korea (NRF) Grant funded by the Korean government (MEST) (no. NRF-2014R1A2A2A01003618) and the Pioneer Research Center Program (Grant no. NRF-2012-0009578).

Author details

Daejoong Kim* and Majid Shahbabaei

*Address all correspondence to: daejoong@sogang.ac.kr

Department of Mechanical Engineering, Sogang University, Seoul, Republic of Korea

References

[1] Cohen-Tanugi D, McGovern RK, Dave SH, Lienhard JH, Grossman JC. Quantifying the potential of ultra-permeable membranes for water desalination. Energy & Environmental Science. 2014;**7**:1134-1141

[2] Fischbein MD, Drndi M. Electron beam nanosculpting of suspended graphene sheets. Applied Physics Letters. 2008;**93**:113107

[3] Nair RR, Wu HA, Jayaram PN, Grigorieva IV, Geim AK. Unimpeded permeation of water through helium-leak-tight graphene-based membranes. Science. 2012;**335**:442-444

[4] Cohen-Tanugi D, Grossman JC. Water desalination across nanoporous graphene. Nano Letters. 2012;**12**:3602-3608

[5] Konatham D, Yu J, Ho TA, Striolo A. Simulation insights for graphene based water desalination membranes. Langmuir. 2013;**29**:11884-11897

[6] Sun C, Boutilier MS, Au H, Poesio P, Bai B, Karnik R, Hadjiconstantinou NG. Mechanisms of molecular permeation through nanoporous graphene membranes. Langmuir. 2014;**30**:675-682

[7] Wang EN, Karnik R. Water desalination graphene cleans up water. Nature Nanotechnology. 2012;**7**:552-554

[8] Koenig SP, Wang L, Pellegrino J, Bunch JS. Selective molecular sieving through porous graphene. Nature Nanotechnology. 2012;**7**:728-732

[9] Kim HW, Yoon HW, Yoon S-M, Yoo BM, Ahn BK, Cho YH, Shin HJ, Yang H, Paik U, Kwon S, et al. Selective gas transport through few-layered graphene and graphene oxide membranes. Science. 2013;**342**:91-95

[10] Li H, Song Z, Zhang X, Huang Y, Li S, Mao Y, Ploehn HJ, Bao Y, Yu M. Ultrathin, molecular-sieving graphene oxide membranes for selective hydrogen separation. Science. 2013;**342**:95-98

[11] Cohen-Tanugi D, Lin L-C, Grossman J. Multilayer nanoporous graphene membranes for water desalination. Nano Letters. 2016;**16**:1027-1033

[12] Shahbabaei M, Kim D. Molecular dynamics simulation of water transport mechanisms through nanoporous boron nitride and graphene multilayers. The Journal of Physical Chemistry B. 2017;**121**:4137-4144

[13] Shahbabaei M, Tang D, Kim D. Simulation insight into water transport mechanisms through multilayer graphene-based membrane. Computational Materials Science. 2017;**128**:87-97

[14] Yoshida H, Bocquet L. Labyrinthine water flow across multilayer graphene-based membranes: Molecular dynamics versus continuum predictions. The Journal of Chemical Physics. 2016;**144**:234701

[15] Dikin DA, Stankovich S, Zimney EJ, Piner RD, Dommett GHB, Evmenenko G, Nguyen ST, Ruoff RS. Preparation and characterization of graphene oxide paper. Nature. 2007;**448**:457-460

[16] Novoselov KS, Falko VI, Colombo L, Gellert PR, Schwab MG, Kim K. A roadmap for graphene. Nature. 2012;**490**:192-200

[17] Joshi RK, Alwarappan S, Yoshimura M, Sahajwalla V, Nishina Y. Graphene oxide: The new membrane material. Applied Materials Today. 2015;**1**:1-12

[18] Yoon HW, Cho YH, Park HB. Graphene-based membranes: Status and prospects. Philosophical Transactions of the Royal Society of London, Series A. 2016;**374**:20150024

[19] Agre P. Aquaporin water channels (noble lecture). Angewandte Chemie International Edition. 2004;**43**:4278-4290

[20] Kumar M, Grzelakowski M, Zilles J, Clark M, Meier W. Highly permeable polymeric membranes based on the incorporation of the functional water channel protein Aquaporin Z. Proceedings of the National Academy of Sciences of the United States of America. 2007;**104**:20719-20724

[21] Kim MJ, McNally B, Murata K, Meller A. Characteristics of solid-state nanometer pores fabricated using a transmission electron microscope. Nanotechnology. 2007;**18**:205302

[22] Gravelle S, Joly L, Detcheverry F, Ybert C, Cottin-Bizonne C, Bocquet L. Optimizing water permeability through the hourglass shape of aquaporins. Proceedings of the National Academy of Sciences. 2013;**110**:16367-16372

[23] Gravelle S, Joly L, Detcheverry F, Ybert C, Bocquet L. Large permeabilities of hourglass nanopores: From hydrodynamics to single file transport. The Journal of Chemical Physics. 2014;**141**:18C526

[24] Shahbabaei M, Kim D. Effect of hourglass-shaped nanopore length on osmotic water transport. Chemical Physics. 2016;**477**:24-31

[25] Su J, Yang K, Guo H. Asymmetric transport of water molecules through a hydrophobic conical channel. RSC Advances. 2014;**4**:40193-40198

[26] Li W, Yan Y, Wang M, Kral P, Dai C, Zhang J. Correlated rectification transport in ultranarrow charged nanocones. Journal of Physical Chemistry Letters. 2017;**8**:435-439

[27] Chinappi M, De Angelis E, Melchionna S, Casciola CM, Succi S, Piva R. Molecular dynamics simulation of ratchet motion in an asymmetric nanochannel. Physical Review Letters. 2006;**97**:144509

[28] Joseph S, Aluru NR. Why are carbon nanotubes fast transporters of water? Nano Letters. 2008;**8**:452-458

[29] Thomas JA, McGaughey AJH. Reassessing fast water transport through carbon nanotubes. Nano Letters. 2008;**8**:2788-2793

[30] Sokhan VP, Nicholson D, Quirke N. Fluid flow in nanopores: Accurate boundary conditions for carbon nanotubes. The Journal of Chemical Physics. 2001;**115**:3878-3887

[31] Cicero G, Grossman JC, Schwegler E, Gygi F, Galli G. Water confined in nanotubes and between graphene sheets: A first principle study. Journal of the American Chemical Society. 2008;**130**:1871-1878

[32] Koga K, Gao GT, Tanaka H, Zeng XC. How does water freeze inside carbon nanotubes. Physica A. 2002;**314**:462

[33] Kolesnikov AI, Zanotti JM, Loong CK, Thiyagarajan P, Moravsky AP, Loutfy RO, Burham CJ. Anomalously soft dynamics of water in a nanotube: A revelation of nanoscale confinement. Physical Review Letters. 2004;**93**:035503

[34] Zangi R, Mark AE. Monolayer ice. Physical Review Letters. 2003;**91**:025502

[35] Zangi R, Mark AE. Bilayer ice and alternate liquid phases of confined water. The Journal of Chemical Physics. 2003;**119**:1694

[36] Hirunsit P, Balbuena PB. Effects of confinement on water structure and dynamics: A molecular simulation study. Journal of Physical Chemistry C. 2007;**111**:1709

[37] Chakraborty S, Kumar H, Dasgupta C, Maiti PK. Confined water: Structure, dynamics, and thermodynamics. Accounts of Chemical Research. 2017;**50**:2139-2146

[38] Zhu FQ, Tajkhorshid E, Schulten K. Theory and simulation of water permeation in Aquaporin-1. Biophysical Journal. 2004;**86**:50-57

[39] Plimpton S. Fast parallel algorithms for short-range molecular dynamics. Journal of Computational Physics. 1995;**117**:1-19

[40] Groot RD, Warren PB. Dissipative particle dynamics: Bridging the gap between atomistic and mesoscopic simulation. The Journal of Chemical Physics. 1997;**107**:4423-4435

[41] Haile JM. Molecular Dynamics Simulation: Elementary Methods. New York: John Wiley & Sons Inc.; 1997

[42] Ryckaert JP, Ciccotti G, Berendsen HJC. Numerical integration of the cartesian equations of motion of a system with constraints: Molecular dynamics of n-alkanes. Journal of Computational Physics. 1977;**23**:327-341

[43] Babu JS, Sathian SP. The role of activation energy and reduced viscosity on the enhancement of water flow through carbon nanotubes. The Journal of Chemical Physics. 2011;**134**:194509

[44] Suk ME, Aluru NR. Molecular and continuum hydrodynamics in graphene nanopores. RCS Advances. 2013;**3**:9365

[45] Wang GJ, Hadjiconstantinou NG. Why are fluid densities so low in carbon nanotubes? Physics of Fluids. 2015;**27**:052006

[46] Suk ME, Aluru NR. Water transport through ultrathin graphene. Journal of Physical Chemistry Letters. 2010;**1**:1590-1594

[47] Ohba T, Kaneko K, Endo M, Hata K, Kanoh H. Rapid water transportation through narrow one-dimensional channels by restricted hydrogen bonds. Langmuir. 2013;**29**:1077

[48] Yoshida H, Mizuno H, Kinjo T, Washizu H, Barrat J-L. Molecular dynamics simulation of electrokinetic flow of an aqueous electrolyte solution in nanochannels. The Journal of Chemical Physics. 2014;**140**:214701

[49] Barati Farimani A, Aluru NR. Spatial diffusion of water in carbon nanotubes: From fickian to ballistic motion. The Journal of Physical Chemistry B. 2011;**115**:12145-12149

[50] Liu Y, Wang Q. Transport behavior of water confined in carbon nanotubes. Physical Review B. 2005;**72**:085420

[51] Bai J, Xing Z, Shan J, Yu H, Duan X. Graphene nanomesh. Nature Nanotechnology. 2010;**5**:190-194

[52] Akhavan O, Ghaderi E. Toxicity of graphene and graphene oxide nanowalls against bacteria. ACS Nano. 2010;**4**:5731-5736

[53] Ning G, Fan Z, Wang G, Gao J, Qian W, Wei F. Gram-scale synthesis of nanomesh graphene with high surface area and its application in supercapacitor electrodes. Chemical Communications. 2011;**47**:5976-5978

[54] Lin Y, Watson KA, Kim JW, Baggett DW, Working DC, Connell JW. Bulk preparation of holey graphene via controlled catalytic oxidation. Nanoscale. 2013;**5**:7814-7824

Chapter 3

Sensors from Electrospun Nanostructures

Liliana Rozemarie Manea and Andrei-Petru Bertea

Additional information is available at the end of the chapter

http://dx.doi.org/10.5772/intechopen.78289

Abstract

Nanotechnology exerts a significant influence on materials science, providing new insights into the design of functional materials. One of the most studied areas of nanotechnologies is that of nanofibres, characterised by high specific volume, chemical activity and volume-dependent physical processes. The most promising method of producing nanofibres with various morphologies and functionalities from different materials is electrospinning, where high voltage is applied between the spinneret and the collector to the charged polymer solution (or melt) to draw polymer filaments. This chapter reviews the main electrospinning techniques for producing nanofibres from polymers, provides an overview on the influence of the spinning solution characteristics, the process parameters and the working environment on the process and highlights the many applications of electrospun nanofibres in the field of sensors. Latest advances in this field and the prospects for obtaining new electrospun nanofibre sensors are discussed.

Keywords: nanostructures, electrospinning, electrospun polymers, electrospinning techniques, sensors

1. Introduction

According to the definition adopted by the National Nanotechnology Initiative, nanotechnology is the manipulation of matter with at least one dimension in the range of 1–100 nanometres [1]. At the moment nanotechnology, an interdisciplinary field having confluence of physics, chemistry and engineering [2] is one of the most attractive areas of science [3], with applications in various domains, which radically effected progress in the field of material science [4]. Nanofibre technology, which comprises the synthesis, processing, fabrication and application of nanoscale fibres [5], is one of the main progressions in nanotechnology [6].

2. Electrospinning: principles and main techniques

Among the technologies that can be used to obtain nanofibres (phase separation, template synthesis, self-assembly, melt blowing) [7], electrospinning is one of the most promising technologies, being commonly used for producing nanosize filaments from both organic polymers and inorganic materials [8], as it is uncomplicated, versatile and capable of producing controlled size fibres with high surface to volume ratio, hence more capable to interrelate with the surrounding environment. In addition, the electrospun nanofibre structures are characterised by complex three-dimensional open porous assembly, facilitating even more the interactions with the adjacent environment. The fact that nanofibres have only one dimension at nanoscale, since the others are at macroscopic one, makes it possible to mingle the advantages of nanostructures (high chemical and biological reactivity and electroactivity) with those of conventional solid membranes, such as comfortable manipulation and easy applicability.

A polymer solution or melt can be electrospun if it is able to carry an electric charge and has enough viscosity to be strained without breaking up into droplets. Up to now, more than 200 types of materials have been electrospun into nanofibres, among which natural and synthetic polymers and many hybrid blends [6].

The principle of electrospinning is quite simple: by applying high voltages electrostatic field (e.g. 10–50 kV) to a polymer solution or a polymer melt, the surface of the fluid elongates and first forms a conical shape, known as the Taylor cone. Then, when the electric voltage reaches a threshold value, the jet of charged liquid overcomes the surface tension, leaves the Taylor cone and is drawn to the collector of different potential, forming a nanofibre net [9].

In addition to the classic electrospinning process, many variants have been developed, capable of producing nanofibres with special features: core/sheath nanofibres [10], nanofibres with hollow structures, nanofibres with porous structures [11] and necklace-like and ribbon nanofibres [12].

The elementary electrospinning setup has four parts: a reservoir of polymer in melt or solution form, a spinneret (in the simplest cases, a syringe), the high voltage power supply and the collector, which acts as a counter electrode (**Figure 1**).

Developments in electrospinning technology have made it possible to electrospin, besides polymers, polymers loaded with nanoparticles and functional molecules, ceramic materials and metal oxides. In addition, there have been developed fibres with new special structures [13].

2.1. Factors that influence the properties of 3D nanostructures obtained by electrospinning

In addition to this basic setup showed in **Figure 1**, there are relatively many other approaches to the process, but in all cases the electrospinning process is influenced by the following three categories of factors: the particularities of the spinning polymer solution/melt, the parameters of the electrospinning process and the environment factors.

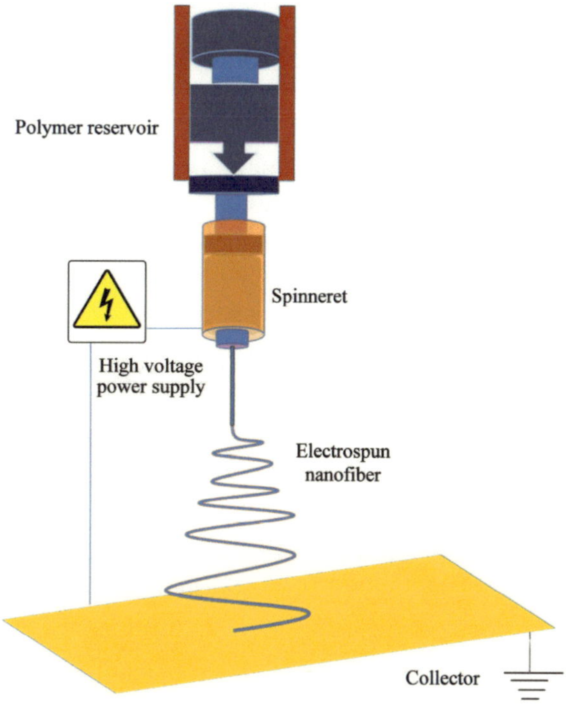

Figure 1. Typical solution electrospinning setup.

2.1.1. Polymer solution characteristics that influence the solution electrospinning process

The main characteristics of the polymer solution with a significant role in the result obtained after electrospinning are concentration, viscosity, conductivity, superficial tension and the volatility characteristics of the solvent.

The solution viscosity is considered to be the dominant variable that determines the fibre diameter. To make electrospinning possible, the viscosity of the electrospun polymer solution must be within a precisely defined range. A too low viscosity leads to interrupted polymer filaments and the appearance of polymer droplets [14], while too high viscosity makes polymer extrusion impossible. The optimum values of viscosity differ according to the molecular weight of the polymer and the solvent used. Of course, there is a close correlation between the concentration of the polymer solution and its viscosity.

An innovative approach to reducing the viscosity of the polymer solution is to apply low-frequency vibrations during the process [15, 16], when vibrations disrupt some of the van der Waals interchain interactions, disordering the polymer chains, which leads to a decrease in the solution viscosity.

Electrospinning fundamentally requires the transfer of electrical charge from the electrode to the polymer droplet at the end point of the syringe needle, so a minimum electrical conductivity in solution is therefore essential for the process; solutions lacking conductivity cannot be electrospun.

A correction of insufficient polymer solution conductivity can be obtained by adding an electrolyte, when the increased number of charges results in an increase in the solution elongation capacity, favouring the formation of smooth, small diameter fibres [17].

Surface tension, strongly influenced by the nature of solvent, is a very important factor in electrospinning; basically, if all the other conditions are established, surface tension controls the upper and lower limits of the range in which the electrospinning can be achieved [18, 19]. Surface tension can be adjusted by varying the polymer/solvent ratio in the spinning solution or by adding surfactant to the solution, which ensures the formation of more uniform fibres [20].

2.1.2. Processing parameters that influence the solution electrospinning process

The applied voltage is one of the most important parameters of the electrospinning process, as it influences directly both the dynamics of fluid flow and the morphological characteristics of the electrospun fibres.

Only when the applied voltage exceeds a certain threshold value, load charged polymer jets are ejected from the Taylor cone. The size of this threshold voltage depends on the nature of the polymer-solvent system [21].

It is generally accepted that an increase in the applied voltage leads to an increase in the deposition rate, thus there is a greater probability of defect formation [22, 23]. The length and diameter of the electrospun fibres decrease with the increase in applied voltage without any change in pore size [20]. For nanofibres produced with low voltage, a uniform morphology with fewer defects and drops is obtained [24].

The distance between the electrode and the collector is important because, if correctly set, can control the morphology and diameters of the nanospun fibres. If the distance is too small, the fibres will not have enough time to solidify before reaching the collector, while if the distance is too long, it is possible to obtain fibre with beds at the surface [25]. A slight change in this distance significantly influences the characteristics of the fibres, the diameter of which becoming smaller as the distance to the collector increases.

The flow rate of the polymer solution within the syringe is an important process parameter; generally, a lower flow rate is recommended to give the polymer solution sufficient time for polarisation. If the flow rate is too high, fibres with many beads are formed due to the insufficient time for the polymer filament to solidify before reaching the collector [26].

2.1.3. Environmental parameters that influence the solution electrospinning process

The influence of the humidity of the environment in which electrospinning takes place is manifested in terms of fibre morphology, deposition orientation and solvent evaporation rate [27]. At very low humidity, a volatile solvent can be dried very quickly, while high humidity helps to discharge static electricity to electrospun fibres. In addition, at high humidity, condensation may occur at the surface of the fibre due to the cooling of the surface of the jet determined by the rapid evaporation of the solvent, and the air flow can interrupt the formation of fibres, causing ruptures [20].

The temperature of the electrospinning environment has a significant influence on the process, as the evaporation rate of the solvent decreases exponentially with the decrease in temperature; when the evaporation process of the solvent becomes slower, the jet takes a longer time to solidify, which can lead to defects in fibre formation [28]. Consequently, temperature control is essential to adjust the evaporation rate of the solvent and the viscosity of the solution [29].

At low atmospheric pressure, the polymer solution in the syringe tends to flow, causing an unstable jet initiation, and at very low pressures, electrospinning cannot be achieved due to direct discharge of electrical charges.

2.1.4. Particularities of melt polymer electrospinning

Melt electrospinning has a number of advantages over solution electrospinning, mainly linked to the absence of solvent, which reduces costs and makes the process more environmentally friendly [30]. In addition, by melt electrospinning, it is possible to transform into nanofibre polymers for which there are no suitable solvents, such as polyolefin or polyethylene terephthalate, or mixtures of polymers for which it is difficult to find an unique solvent for all the components [31, 32]. Unlike the solution electrospinning, where frequently a nonwoven fibrous mat is obtained, the melt electrospinning can produce filaments which can be used in knitting or weaving processes [33]. In this case, the melt electrospinning process (known as melt electrospinning writing) uses a moving collector, which exerts a translational movement sufficiently fast for the rectilinear deposition of the polymer jet [34].

In addition to these advantages, there are difficulties arising from the particular equipment required (**Figure 2**) and the high viscosity and low electrical conductivity of melt polymers.

The main feature of the polymer melt that influences electrospinning is viscosity. As the viscosity of the melt is at least one order of magnitude greater than that of the polymer solutions, it is essential to reduce it. This is usually done by raising the temperature (without degrading the polymer) or by adding additives such as cationic surfactants [35, 36].

Another parameter related to the polymer melt is the molecular weight of the polymer, which influences the diameter of the obtained nanofibres. When the molecular weight of the electrospun polymer is higher, the Melt flow index is lower and the diameter of the fibres is bigger. Because in the melt electrospinning, the viscosity cannot be controlled by adjusting the polymer/solvent ratio, like in the solution electrospinning, it is essential to choose a suitable molecular weight of the polymer to achieve a stable and reproducible electrospinning process [33]. It is believed that the optimal range of molecular weights for melt electrospinning is substantially lower than that characteristic of solution electrospinning. The process is also influenced by the stereoregularity of the polymer, the isotactic polymers producing finer fibres than the atactic ones [32].

The conductivity of the polymer melt decisively influences the stability of the extruded polymer jet. A too high conductivity determines the jet's instability, while a too small one will cause reduced electrostatic traction forces. It is considered that average conductivity values in the range of 10^{-6}–10^{-8} S/m prerequisites for a stable electrospinning process [33].

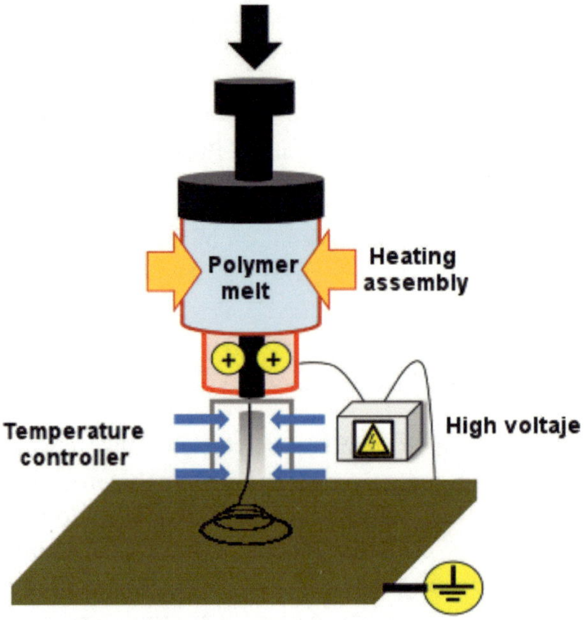

Figure 2. Melt electrospinning setup.

In what regards the processing parameters that influence the melt electrospinning process, it can be seen that, unlike the solution electrospinning, there is a direct proportionality relationship between the fibre diameter and the melt polymer flow rate.

Concerning the effect of the applied voltage on the diameter of the fibres, information is contradictory, but the overall conclusion is that for each polymer there is an optimum voltage range [37–40].

The distance between the spinneret and the collector is relatively lower in melt electrospinning than in the solution process (usually 3–5 cm). This distance influences the cooling process of the polymer melt jet and the shape of the fibres. Flat fibres may appear at smaller spinneret to collector distances, while larger distances affect the accuracy of depositing filaments on the collector in the melt electrospinning writing process [34].

3. Electrospun sensors

Sensors, which convert energy to detect concentration changes of a specific substance and communicate the information in the form of an electrical or optical signal [41], have gained special importance in recent decades, due to their numerous applications in areas such as monitoring environmental factors, health and wellbeing or detection of dangerous vapours [42].

A high specific surface area and a very porous structure are vital for high sensitivity and quick response of a sensor [31]. This is why electrospun nanomaterials, having large specific surface area and high and tunable porosity [43], have found frequent use in the field of sensors of many types, such as chemiresistive, optic (fluorescent and colorimetric), acoustic wave (piezoelectric) or photoelectric [13]. Electrospun nanostructured sensors display faster adsorption and minimised bulk effects when compared to the conventional sensors [44].

3.1. Chemiresistive electrospun sensors

Chemiresistive sensors, which work by measuring resistance variation when in contact with the substance to be detected, have a high miniaturisation potential, making them applicable to portable devices. Semiconductive polymers, metal oxides, metals and conjugated compounds are used to obtain such sensors [42, 45].

Polymers that exhibit conductivity, such as polyaniline (PANI), polypyrrole (PPY) and poly (3, 4-ethylenedioxythiophene) (PEDOT), are used as gas sensors [46]. For example, PEDOT, insoluble in pure state, becomes dispersible in water after doping with poly (styrenesulfonate) (PSS) and can be electrospun, the obtained nanofibres being used for the detection of inorganic compounds such as nitrogen oxide and ammonia and organic substances (ethanol, methanol and acetone) by measuring resistance variation [47, 48].

Polyaniline (PAni) doped with (+)-camphor-10-sulfonic acid (HCSA) has been electrospun and proved to be excellent ammonia sensors, while the electrospun fibres obtained from undoped PAni exhibited very good nitrogen oxide detection capabilities [49]. Polyaniline electrospun nanofibres also proved to be effective as moisture sensor, when the response to changes was given by impedance variation. It was found that the sensor exhibited a rapid, reversible and very sensitive change in impedance as a function of adsorbed water molecules [50]. A sensor from electrospun nanofibres from polyaniline/polyethylene oxide (PAni-PEO) doped with 10-camphosulfonic acid (HCSA) and deposited on a pair of gold electrodes modifies resistance, having a 0.5 ppm detection threshold for ammonia [51].

Chemiresistive sensors from PAN/PAni electrospun fibres showed good sensitivity to NH_3 gas at room temperature and this sensitivity increased with increasing gas concentration [52].

Ammonia sensors from quartz crystal microbalance coated with electrospun polyacrylic acid (PAA) membranes showed high sensitivity at low ppb level, but this sensitivity depended on the morphology of the nanofibres, the load of the quartz crystal microbalance with the electrospun membrane and the relative humidity [53].

Another ammonia sensor was obtained by depositing a P-type conductive PAni onto the surface of n-type semiconductive TiO_2 electrospun fibres. The sensitivity of this sensor substantially increases with the increase of NH_3 concentration [54].

Electrospun nanofibres from PANI and PANI/ZnO were used to detect HCl and NH_3 vapours at room temperature, recording a decrease in the resistance of the sensors when exposed to HCl vapour, respectively an increase in resistance when exposure to NH_3 vapours. The PAni/ZnO sensor has a high sensitivity response at room temperature with better repeatability compared to that of the pure PANI sensor [55].

An ammonia sensor based on a single PAni nanofibre demonstrated a very low detection value (under 1 ppm) and a response time under 10 s [56].

Electrospun nanofibres of poly (o-anisidine)-polystyrene doped with (+)-camphor-10-sulfonic acid proved to be effective in water and ethanol determination, showing a high sensitivity and rapid response and recovery [57].

Electrospun conductive polypyrrole nanofibres coating a copper interdigital electrode has been used to produce an aliphatic amines gas sensor with low detection limit (0.42 ppm for *n*-butylamine), quick response and good repeatability at temperatures between 90 and 200°C [58]. PAni/poly (vinyl pyrrolidone) nanofibres containing urease have been used as NO_2 sensors, showing a significant conductivity increase in the presence of levels of NO_2 in the range of 1–7 ppm [59].

Electrospun nanofibres of Al-SnO_2/PAni were used as hydrogen sensors, having high sensitivity at low temperature (48°C) [60], while sensors made of PAni/TiO_2:SnO_2 nanofibres placed onto an epoxy glass substrate detected hydrogen at even lower temperature (27°C) showing good sensitivity [61].

Electrospun polyamide nanofibrous membranes have been proved to be effective as glucose biosensors, displaying a sensitivity of 1.11 μA/mM and a detection limit of 2.5×10^{-6} M [62].

TiO_2 fibre mats obtained by electrospinning from a dimethyl formamide solution of poly (vinyl acetate) showed high sensitivity to NO_2 and H_2 with reversible response and a response times of the order of 1 min [63].

Sensors for humidity and KCl were obtained from electrospun poly (lactic acid)/polyaniline fibres deposited onto interdigitated microelectrodes. The sensitivity of the sensor depends on the poly (lactic acid)/polyaniline ratio in the blend [64].

Electrospun nanofibre sensors have found applications in the medical field as well. A sensor composed of a electrospun conductive nanofibres mat with polyacrylic acid grafted on its surface and a sensing element made of conducting polymer with covalently attached oligonucleotide probes is effective in determining the non-Hodgkin's lymphoma gene, with a detection limit of 1 aM (1 × 10–18 mol/L) and high selectivity [65].

Sensing electrodes for catechol detection were obtained by electrospinning a PAni solution with dispersed multiwall carbon nanotubes followed by the reduction of the amino groups and the addition of polyphenol oxidase on the nanofibres [66].

3.2. Optical electrospun sensors (colorimetric and fluorescent)

Optical sensors work by altering the colour or intensity of their fluorescence by the analyte [42]. These sensors, which are usually based on a technology that allows miniaturisation, low cost and in-situ usage, show high sensitivity and selective response towards various analytes [67].

Compared to other methods of obtaining nanofibres, electrospinning provides nanofibres for optical sensors, which are easy to manufacture and operate, relatively low-priced and with customizable properties, such as diameter, morphology and porosity [68].

There are three ways to obtain optical sensors by electrospinning: inserting chromophores into a transparent polymer without optical properties; using as substrate polymers having the ability to absorb/emit light and the functionalization of the electrospun polymer with active optical nanosystems [42].

Electrospun core-shell nanofibre with polymethyl methacrylate shell and a core made of phase change thermochromic material show thermochromic phase change comportment and can be used in detecting foreign materials in a specific environment [69].

A colorimetric sensor made of electrospun porphyrinated polyimide (PPI) nanofibrous membrane proved to be effective in hydrogen chloride detection, showing high sensitivity and fast response time. When exposed cu HCl vapours, the polymer changes its colour from pink to green, as a result of the protonation of the neutral porphyrin [70], and the colour intensity increases as the concentration of HCl is higher.

Colorimetric strip sensors to detect uranyl (UO_{22}^{+}) have been produced using electrospun cellulose acetate nanofibre mats that incorporated 2-(5-Bromo-2-pyridylazo)-5-(diethylamino) phenol. The colour in these sensors changes from yellow to purple in the presence of uranyl at pH 6.0, and the detection limit can extend to 50 ppb, with very low interference ability in the presence of other metal ions [71].

Colorimetric sensors have been produced incorporating gold nanoparticles into electrospun polystyrene nanofibres. These sensors, which do not employ any complex instrumentation, showed very good response towards oestrogenic compounds such as 17β-estradiol, with high sensitivity [72].

Optically active electrospun nanofibres can be used as sensors to detect heavy metals using colorimetry and fluorescence methods [68].

Colorimetric sensors based on nanofibres electrospun from a mixture of 10, 12-pentacosadiynoic acid and poly (ethylene oxide) and polyurethane can be used to detect *Escherichia coli* [73].

Membranes from electrospun polyamide 66/cobalt chloride proved to be effective in determining humidity, due to the colour change property of cobalt salt, which turns from blue to pink as the relative humidity increases over 12.4%. The sensor has high sensitivity, fast response time, good reproducibility and long-time stability [74].

Mats from electrospun cellulose acetate fibres doped with the chromogenic and fluorogenic amine-reactive blue dye Py-1 act as colorimetric sensors detecting biogenic amines [75].

A nanofibre-based fluorescent sensor for Ni^{2+} determination, even in the presence of other competing metal ions, was obtained by electrospinning pyridylazo-2-naphthol-poly(acrylic acid) polymer functionalized with 1,10-carbonyldiimidazole and 1,8-diazabicyclo[5.4.0]undec-7-ene [76]. The sensor showed high sensitivity and selectivity in its fluorescence towards Ni^{2+}, and the detection was simple, rapid and selective, without any need of additional sample handling phases.

Optical fluorescence sensors based on poly (acrylic acid)-poly (pyrenemethanol) polymers have proven effective for the detection of metal ions (Fe^{3+} and Hg^{2+}) and 2,4-dinitrotoluene (DNT), showing high sensitivity due to the high-volume surface ratio of the nanofibre membrane structures [77].

A fluorescence sensor based on a pentiptycene conjugated polymer was used for Copper (II) cation detection, with high selectivity and sensitivity [78].

A fluorescent film sensor made of electrospun nanofibres from a copolymer of vinyl naphthalimide and methyl methacrylate was used for Cu^{2+} detection, showing high sensitivity (a detection limit of 20×10^{-6} M Cu^{2+}) [79].

A picric acid fluorescent sensor was obtained by electrospinning a solution of 5-(N-carbazole styryl)-1, 3-dimethyl-barbituric acid and polystyrene in dimethyl formaldehyde/tetrahydrofuran upon an amino-functionalized glass. The emission intensity of the fluorescent sensor is not significantly influenced by common interferents and the fluorescent nanofibres can be regenerated [80].

Fluorescence sensors for detecting traces of toluene vapours have been produced by embedding fluorescent CdTe quantum dots in electrospun polyvinyl alcohol nanofibres, showing sensitivity and fast time response [81].

A fluorescence sensor for Cu^{2}+ sensing, which displays good sensitivity and selectivity, was obtained from electrospun rhodamine dye-doped polyester nanofibres [82].

A fluorescent sensor for the rapid detection of bacteria was made from electrospun membranes of a boronic acid copolymer, poly (4-vinylphenylboronic acid-co-2-(dimethylamino) ethyl methacrylate-co-n-butyl methacrylate), showing high affinity and towards both Gram-negative and Gram-positive bacteria [83].

Optical sensors based on electrospun polyvinyl alcohol nanofibres with embedded nanocomposite cerium oxide nanoparticles can be used to detect radicals in solutions. These nanocomposite nanoparticles are fluorescent with observable emission under near-UV excitation [84].

3.3. Other electrospun sensors

Besides the chemiresistive and optical sensors, other categories of electrospun nanofibre sensors have been developed.

An electrospun nanofibre piezoelectric sensor made of polyvinylidene fluoride (PVDF) nanofibre nonwoven webs proved to be effective in acoustic sensing for various applications [85].

A humidity sensor was obtained by depositing polyaniline composite nanofibres on surface acoustic wave resonator, demonstrating very high sensitivity, fast response and good sensing linearity [86].

Sensors based on gas adsorption on electrospun PAni fibre mats have been used to determine carbon dioxide. The process is reversible, and the adsorption process is fast enough to make quick detection possible [87].

Surface acoustic wave sensors are piezoresistive sensors [88]. Such a sensor based on creased polypyrrole film with electrospun polyvinyl alcohol nanowires as spacer showed high sensitivity, low detection limit and good stability [89]. Other piezoresistive sensors have been obtained from polyvinylidene fluoride and its blend with polyoctafluoropentyl acrylate [90].

4. Conclusions

Due to their small size and high surface-to-volume ratio, many sensors based on electrospun nanofibres have been used lately for analyte detection, proving great sensitivity and very rapid response time, all in elevated stability conditions. Nanotechnology is constantly developing towards new challenges in many areas including sensors. In future developments, it can be expected that the problems that affect the efficacy of electrospun nanofibre sensors are to be addressed and solved, beginning with a more accurate control of the process in order to achieve better control over the size and morphology of the nanofibres and continuing with increasing pore uniformity for more sensitivity and improving the reversibility of the sensors.

Acknowledgements

This work was supported by a grant from the Romanian National Authority for Scientific Research and Innovation.

Author details

Liliana Rozemarie Manea* and Andrei-Petru Bertea

*Address all correspondence to: andrei_bertea@yahoo.co.uk

Technical University of Iasi, Textile, Leather and Industrial Management Faculty, Iasi, Romania

References

[1] Meddocs. Electrospun Nanofibres as 3D-Structures for Nanomedicine. [Internet]. 2018. Available from: http://www.meddocsonline.org/journal-of-nanomedicine/electrospun-nanofibres-as-3d-structures-for-nanomedicine.pdf [Accessed: March 23, 2018]

[2] Kumar C, Poornachandra Y, Pombala S. Therapeutic nanomaterials: From a drug delivery perspective. In: Andronescu E, Grumezescu AM, editors. Nanostructures for Drug Delivery. Amsterdam: Elsevier Inc.; 2017. pp. 1-35. DOI: 10.1007/978-0-323-46143-6

[3] Bagheri S, Amiri IS, Yousefi AT. Nanocomposites in Electrochemical Sensors. London: Taylor & Francis Group; 2017. p. 3

[4] Afshari M. Introduction. In: Afshari M, editor. Electrospun Nanofibres. Cambridge: Woodhead Publishing in Association with The Textile Institute; 2017. pp. 1-8. DOI: 10.1016/B978-0-08-100907-9.00002-7

[5] Ko F, Wan Y. Introduction to Nanofibre Materials. Cambridge: University Printing House; 2014. p. 128. DOI: 10.1007/978-0-521-87983-5

[6] Mohammadzadehmoghadam S, Dong Y, Barbhuiya S, Guo L, Liu D, Umer R, Qi X, Tang Y. Electrospinning: Current status and future trends. In: Fakirov S, editor. Nano-size Polymers. Switzerland: Springer International Publishing; 2016. pp. 89-154. DOI: 10.1007/978-3-319-39715-3_4

[7] Mahenderkar N, Prabhu R, Kumar A. Nanocomposites potential for aero applications. In: Eswara Prasad N, Wanhill J, editors. Aerospace Materials and Material Technologies. Vol. 1. Aerospace Materials. Singapore: Springer Science+Business Media; 2017. pp. 391-411. DOI: 10.1007/978-981-10-2134-3

[8] Baji A, Mai Y-W. Engineering ceramic fiber nanostructures through polymer-mediated electrospinning. In: Lin Z, Yang Y, Zhang A, editors. Polymer-Engineered Nanostructures for Advanced Energy Applications. Cham: Springer International Publishing AG; 2017. pp. 3-30. DOI: 10.1007/978-3-319-57003-7

[9] Li X, Zhang Y, Li H, Chen H, Ding Y, Yang W. Effect of oriented fibre membrane fabricated via needleless melt electrospinning on water filtration efficiency. Desalination. 2014;**344**:266-273. DOI: 10.1016/j.desal.2014.04.003

[10] Yu D-G, Zhu L-M, Branford-White C, Yang J-H, Wang X, Li Y, Qian W. Solid dispersions in the form of electrospun core-sheath nanofibers. Journal of Nanomedicine. 2011;**6**:3271-3280. DOI: 10.2147/IJN.S27468

[11] Khajavi R, Abbasipour M. Electrospinning as a versatile method for fabricating coreshell, hollow and porous nanofibers. Scientia Iranica. 2012;**19**:2029-2034. DOI: 10.1016/j.scient.2012.10.037

[12] Koombhongse S, Liu W, Reneker D. Flat ribbons and other shapes by electrospinning. Journal of Polymer Science Part B Polymer Physics. 2001;**39**:2598-2606. DOI: 10.1002/polb.10015

[13] Huang J, You T. Electrospun nanofibres: From rational design, fabrication to electrochemical sensing applications. In: Maguire R, editor. Advances in Nanofibres. Rijeka: IN-TECH; 2017. pp. 35-83. DOI: 10.5772/57099

[14] Abd Razak SI, Wahab IF, Fadil F, Dahli F, Khudzari A, Adeli H. A review of electrospun conductive polyaniline based nanofibre composites and blends: Processing features, applications, and future directions. Advances in Materials Science and Engineering. 2015;**2015**:1-18. DOI: 10.1155/2015/356286

[15] He JH, Wan YQ, Yu JY. Application of vibration technology to polymer electrospinning. International Journal of Nonlinear Sciences and Numerical Simulation. 2004;**5**:253-262. DOI: 10.1515/ijnsns.2004.5.3.253

[16] Zohoori S, Latifi M, Davodiroknabadi A, Mirjalili M. Vibration electrospinning of polyamide-66/multiwall carbon nanotube nanocomposite: Introducing electrically conductive, ultraviolet blocking and antibacterial properties. Polish Journal of Chemical Technology. 2017;**19**:56-60. DOI: 10.1515/pjct-2017-0049

[17] Haghi A K, Zaikov G E. Electrospinning of nanofibres: From introduction to application. Moscow: Nova Science Publishers; 2011. 175 p. DOI: 10.1201/b12229

[18] Li Z, Wang C. One-Dimensional Nanostructures Electrospinning Technique and Unique Nanofibres. Berlin: Springer; 2013. p. 141. DOI: 10.1007/978-3-642-36427-3

[19] Pillay V, Dott C, Choonara C, Tyagi C, Tomar L, Kumar P, du Toit L, Ndesendo V. A review of the effect of processing variables on the fabrication of electrospun nanofibres for drug delivery applications. Journal of Nanomaterials. 2013;**2013**:1-22. DOI: 10.1155/2013/789289

[20] Bhardwaj N, Kundu S. Electrospinning: A fascinating fibre fabrication technique. Biotechnology Advances. 2010;**28**:325-347. DOI: 10.1016/j.biotechadv.2010.01.004

[21] Haider A, Haider S, Kang I-K. A comprehensive review summarizing the effect of electrospinning parameters and potential applications of nanofibres in biomedical and biotechnology. Arabian Journal of Chemistry. 2015;**2015**:1-24. DOI: 10.1016/j.arabjc.2015.11.015

[22] Ramakrishna S, Fujihara K, Teo WE, Lim TC, Ma Z. An Introduction to Electrospinning and Nanofibres. London: World Scientific Publishing Co. p. 382. DOI: 10.1007/981-256-415-2

[23] Sill T, von Recum H. Electrospinning: Applications in drug delivery and tissue engineering. Biomaterials. 2008;**29**:1989-2006. DOI: 10.1016/j.biomaterials.2008.01.011

[24] Tan S-H, Inai R, Kotaki M, Ramakrishna S. Systematic parameter study for ultra-fine fiber fabrication via electrospinning process. Polymer. 2005;**46**:6128-6134. DOI: 10.1016/j.polymer.2005.05.068

[25] Ki CS, Baek DH, Gang KD, Lee KH, Um IC, Park YH. Characterization of gelatin nanofibre prepared from gelatin-formic acid solution. Polymer. 2005;**46**:5094-5102. DOI: 10.1016/j.polymer.2005.04.040

[26] Zargham S, Bazgir S, Tavakoli A, Rashidi A, Damerchely R. The effect of flow rate on morphology and deposition area of electrospun nylon 6 nanofibre. Journal of Engineered Fibers and Fabrics. 2012;**7**:42-49. DOI: 10.1016/j.eng.f.2012. 08.056

[27] Burke L.D, Blackwood K.A, Volpato F. Reproducibility and robustness in electrospinning with a view to medical device manufacturing. In: Almodovar J, ditor. Electrospun Biomaterials and Related Technologies. Cham: Springer International Publishing AG; 2017. p. 1-20. DOI 10.1007/978-3-319-70049-6

[28] De Vrieze S, Van Camp T, Nelvig A, Hagström B, Westbroek P, De Clerck K. The effect of temperature and humidity on electrospinning. Journal of Materials Science. 2009;**44**:1357-1362. DOI: 10.1007/s10853-008-3010-6

[29] Corradini E, Schroeder Curti P, da Silva Gonzalez R, Pereira A, Dragunski D, Martins A, Curti Muniz E. Preparation of polymeric mats through electrospinning for technological uses. In: Longo E, de Almeida La Porta F, editors. Recent Advances in Complex Functional Materials. From Design to Application. Cham: Springer International Publishing AG; 2017. pp. 83-130. DOI: 10.1007/978-3-319-53898-3

[30] Muerza-Cascante M, Haylock D, Hutmacher D, Dalton P. Melt electrospinning and its technologization in tissue engineering. Tissue Engineering: Part B. 2015;**21**:187-202. DOI: 10.1089/ten.teb.2014.0347

[31] Nayak R, Padhye R, Arnold L. Melt-electrospinning of nanofibres. In: Afshari M, editor. Electrospun Nanofibres. Cambridge: Woodhead Publishing in Association with The Textile Institute; 2017. pp. 11-40. DOI: 10.1016/B978-0-08-100907-9.00002-7

[32] Nayak R. Polypropylene Nanofibres—Melt Electrospinning Versus Meltblowing. Cham: Springer International Publishing AG; 2017. p. 190. DOI: 10.1007/978-3-319-61458-8

[33] Brown T, Dalton P, Hutmacher D. Melt electrospinning today: An opportune time for an emerging polymer process. Progress in Polymer Science. 2016;**56**:116-166. DOI: 10.1016/j.progpolymsci.2016.01.001

[34] Dalton P, Muerza-Cascante L, Hutmacher D. Design and fabrication of scaffolds via melt electrospinning for applications in tissue engineering. In: Mitchell G, editor. Electrospinning—Principles, Practice and Possibilities. Cambridge: The Royal Society of Chemistry; 2015. pp. 100-120. DOI: 10.1007/978-1-84973-556-8

[35] Lin T, Wang H, Wang X. The charge effect of cationic surfactants on the elimination of fibre beads in the electrospinning of polystyrene. Nanotechnology. 2004;**15**:1375-1381. DOI: 10.1088/0957-4484/15/9/044

[36] Chen Z, He J, Zhao F, Liu Y, Liu Y, Yuan H. Effect of polar additives on melt electrospinning of non-polar polypropylene. Journal of the Serbian Chemical Society. 2014;**79**:587-596. DOI: 10.2298/JSC130702150Z

[37] Mota C, Puppi D, Gazzarri M, Bártolo P, Chiellini F. Melt electrospinning writing of three-dimensional star poly(ϵ-caprolactone) scaffolds. Polymer International. 2013;**62**:893-900. DOI: 10.1002/pi.4509

[38] Fang J, Zhang L, Sutton D, Wang X, Lin T. Needleless melt-electrospinning of polypropylene nanofibres. Journal of Nanomaterials. 2012;**2-3**:1-11. DOI: 10.1155/2012/382639

[39] Shimada N, Ogata N, Nakane K, Ogihara T. Spot laser melt electrospinning of a fibre bundle composed of poly(lactide)/poly(ethylene-co-vinyl alcohol) pie wedge fibres. Journal of Applied Polymer Science. 2012;**125**:E384-E389. DOI: 10.1002/app.36820

[40] Liu Y, Liu Z, Deng L, Wang K, Yang W. Effect of different factors on falling process of melt electrospinning jet. Materials Science Forum. 2013;**745-746**:407-411. DOI: 10.4028/www.scientific.net/MSF.745-746.407

[41] Qin X. Coaxial electrospinning of nanofibres. In: Afshari M, editor. Electrospun Nanofibres. Cambridge: Woodhead Publishing in Association with The Textile Institute; 2017. pp. 41-71. DOI: 10.1016/B978-0-08-100907-9.00002-7

[42] Choi S-J, Persano L, Camposeo A, Jang J-S, Koo W-T, Kim S-J, Cho H-J, Kim I-D, Pisignano D. Electrospun nanostructures for high performance chemiresistive and optical sensors. Macromolecular Materials and Engineering. 2017;**8**:1-37. DOI: 10.1002/mame.201600569

[43] El-Moghazy AY, Soliman EA, Ibrahim HZ, Marty J-L, Istamboulie G, Noguer T. Biosensor based on electrospun blended chitosan-poly(vinyl alcohol) nanofibrous enzymatically sensitized membranes for pirimiphos-methyl detection in olive oil. Talanta. 2016;**155**:258-264. DOI: 10.1016/j.talanta.2016.04.018

[44] Macagnano A, Zampetti E, Kny E, editors. Electrospinning for High Performance Sensors. Chen: Springer International Publishing; 2015. p. 340. DOI: 10.1007/978-3-319-14406-1

[45] Ziadan KM. Conducting polymers application. In: Gomes A, editor. New Polymers for Special Applications. Rijeka: Intech; 2012. pp. 3-24. DOI: 10.5772/3345

[46] Sołoducho J, Cabaj J. Conducting polymers in sensor design. In: Yılmaz F, editor. Conducting Polymers. Rijeka: Intech; 2016. pp. 27-48. DOI: 10.5772/61723

[47] Pinto NJ, Rivera D, Melendez A, Ramos I, Lim JH, Johnson A. Electrical response of electrospun PEDOT-PSSA nanofibres to organic and inorganic gases. Sensors and Actuators B: Chemical. 2011;**156**:849-853. DOI: 10.1016/j.snb.2011.02.053

[48] Zhang HD, Yan X, Zhang Z-H, Yu G-F, Han W-P, Zhang J-C, Long Y-Z. Electrospun PEDOT:PSS/PVP nanofibres for CO gas sensing with quartz crystal microbalance technique. International Journal of Polymer Science. 2016;**2016**:1-6. DOI: 10.1155/2016/3021353

[49] Zhang Y, Kim J, Chen D, Tuller H, Rutledge G. Electrospun polyaniline fibers as highly sensitive room temperature chemiresistive sensors for ammonia and nitrogen dioxide gases. Advanced Functional Materials. 2014;(25):4005-4014. DOI: 10.1002/adfm.2014-00185

[50] Lin Q, Li Y, Yang M. Highly sensitive and ultrafast response surface acoustic wave humidity sensor based on electrospun polyaniline/poly(vinyl butyral) nanofibers. Analytica Chimica Acta. 2012;**748**:73-80. DOI: 10.1016/j.aca.2012.08.041

[51] Liu HQ, Kameoka J, Czaplewski DA, Craighead HG. Polymeric nanowire chemical sensor. Nano Letters. 2004;**4**:671-675. DOI: 10.1021/nl049826f

[52] Anwane R, Kondawar S, Late D. Bessel's polynomial fitting for electrospun polyacrylonitrile/polyaniline blend nanofibers based ammonia sensor. Materials Letters. 2018;**221**:70-73. DOI: 10.1016/j.matlet.2018.03.075

[53] Ding B, Yamazaki M, Shiratori S. Electrospun fibrous polyacrylic acid membrane-based gas sensors. Sensors and Actuators B. 2015;**106**:477-483. DOI: 10.1016/j.snb.2004.09.010

[54] Pandey S. Highly sensitive and selective chemiresistor gas/vapor sensors based on polyaniline nanocomposite: A comprehensive review. Journal of Science: Advanced Materials and Devices. 2016;**1**:431-453. DOI: 10.1016/j.jsamd.2016.10.005

[55] Kondawar S, Patil P, Agrawal S. Chemical vapour sensing properties of electrospun nanofibres of polyaniline/ZnO nanocomposites. Advanced Functional Materials. 2014; **5**:389-395. DOI: 10.5185/amlett.2014.amwc.1037

[56] Chen D, Lei S, Chen Y. A single polyaniline nanofibre field effect transistor and its gas sensing mechanisms. Sensors. 2011;**11**:6509-6516. DOI: 10.3390/s110706509

[57] Aussawasathien D, Sahasithiwat S, Menbangpung L, Teerawattananon C. Poly(o-anisidine)-polystyrene composite fibres via electrospinning process: Surface morphology and chemical vapor sensing. Sensors and actuators. B: Chemical. 2011;**151**:341-350. DOI: 10.1016/j.snb.2010.07.048

[58] Akbarinejad A, Ghoorchian A, Kamalabadi M, Alizadeh N. Electrospun soluble conductive polypyrrole nanoparticles for fabrication of highly selective *n*-butylamine gas sensor. Sensors and Actuators B. 2016;**236**:99-108. DOI: 10.1016/j.snb.2016.05.034

[59] Manea LR, Bertea AP, Hristian L, Cramariuc R, Cramariuc O. Electrospun based polyaniline sensors—A review. Materials Science and Engineering. 2017;**209**:1-10. DOI: 10.1088/1757-899X/209/1/012063

[60] Sharma H, Jamkar D, Kondawar S. Electrospun nanofibres of conducting polyaniline/Al-SnO_2 composites for hydrogen sensing applications. Procedia Materials Science. 2015;**10**:186-194. DOI: 10.1016/j.mspro.2015.06.040

[61] Nasirian S, Moghaddam HM. Polyaniline assisted by TiO_2:SnO_2 nanoparticles as a hydrogen gas sensor at environmental conditions. Applied Surface Science. 2015;**328**:395-404. DOI: 10.1016/j.apsusc.2014.12.051

[62] Scampicchio M, Arecchia A, Lawrence N, Mannino S. Nylon nanofibrous membrane for mediated glucose biosensing. Sensors and Actuators B: Chemical. 2010;**145**:394-397. DOI: 10.1016/j.snb.2009.12.042

[63] Kim ID, Rothschild D, Tuller H, Kim DY, Jo SM. Electrospun TiO_2 nanofibers for gas sensing applications. In: Technical Proceedings of the 2007 Clean Technology Conference and Trade Show Cleantech. 14-17 May 2007; Austin. pp. 334-338

[64] Picciani P, Medeiros E, Venancio E, Soares B, Mattoso L. Development of electrospun polyaniline/poly(lactic acid) nanofiber-based sensors. Sensor Letters. 2016;**14**:1-8. DOI: 10.1166/sl.2016.3740

[65] Kerr-Phillips T, Aydemir N, Chan E, Barker D, Malmström J, Plesse C, Travas-Sejdic J. Conducting electrospun fibres with polyanionic grafts as highly selective, label-free, electrochemical biosensor with a low detection limit for non-Hodgkin lymphoma gene. Biosensors and Bioelectronics. 2018;**100**:549-555. DOI: 10.1016/j.bios.2017.09.042

[66] Sapountzi S, Braiek M, Chateaux J-F, Jaffrezic-Renault N, Lagarde F. Recent advances in electrospun nanofiber interfaces for biosensing devices. Sensors. 2017;**17**:1887-1915. DOI: 10.3390/s17081887

[67] Piriya A, Joseph P, Daniel K, Lakshmanan S, Kinoshita T, Muthusamy S. Colorimetric sensors for rapid detection of various analytes. Materials Science and Engineering C. 2017;**78**:1231-1245. DOI: 10.1016/j.msec.2017.05.018

[68] Terra I, Mercante L, Andre R, Correa D. Fluorescent and colorimetric electrospun nanofibres for heavy-metal sensing. Biosensors. 2017;**7**:61-75. DOI: 10.3390/bios7040061

[69] Li F, Zhao Y, Wang S, Han D, Jiang L, Song Y. Thermochromic core-shell nanofibres fabricated by melt coaxial electrospinning. The Journal of Applied Polymer Science. 2009;**112**:269-274. DOI: 10.1002/app.29384

[70] Lv Y-Y, Wu J, Xu ZK. Colorimetric and fluorescent sensor constructing from the nanofibrous membrane of porphyrinated polyimide for the detection of hydrogen chloride gas. Sensors and Actuators B. 2010;**148**:233-239. DOI: 10.1016/j.snb.2010.05.029

[71] Hu L, Yan X-W, Li Q, Zhang X-J, Shan D. Br-PADAP embedded in cellulose acetate electrospun nanofibers: Colorimetric sensor strips for visual uranyl recognition. Journal of Hazardous Materials. 2017;**329**:205-210. DOI: 10.1016/j.jhazmat.2017.01.038

[72] Pule B, Degni S, Torto N. Electrospun fibre colorimetric probe based on gold nanoparticles for on-site detection of 17β-estradiol associated with dairy farming effluents. Water SA. 2015;**41**:27-32. DOI: 10.4314/wsa.v41i1.5

[73] Yapor J, Alharby A, Gentry-Weeks C, Reynolds M, Alam M, Li Y. Polydiacetylene nanofiber composites as a colorimetric sensor responding to *Escherichia coli* and pH. ACS Omega. 2017;**2**:7334-7342. DOI: 10.1021/acsomega.7b01136

[74] You M-H, Yan X, Zhang J, Wang X-X, He X-X, M1 Y, Ning X, Long Y-Z. Colorimetric humidity sensors based on electrospun polyamide/$CoCl_2$ nanofibrous membranes. Nanoscale Research Letters. 2017;**12**:360-367. DOI: 10.1186/s11671-017-2139-0

[75] Yurova N, Danchuk A, Mobarez S, Wongkaew N, Rusanova T, Baeumner A, Duerkop A. Functional electrospun nanofibers for multimodal sensitive detection of biogenic amines in food via a simple dipstick assay. Analytical and Bioanalytical Chemistry. 2018;**410**:1111-1121. DOI: 10.1007/s00216-017-0696-9

[76] Adewuyi S, Ondigo D, Zugle R, Tshentu Z, Nyokong T, Torto N. A highly selective and sensitive pyridylazo-2-naphthol-poly(acrylic acid) functionalized electrospun nanofibre fluorescence "turn-off" chemosensory system for Ni^{2+}. Analytical Methods. 2012;**4**:1729-1735. DOI: 10.1039/c2ay25182e

[77] Wang X, Drew C, Lee S-H, Senecal K, Kumar J, Samuelson L. Electrospun nanofibrous membranes for highly sensitive optical sensors. Nano Letters. 2002;**2**:1273-1275. DOI: 10.1021/nl020216u

[78] Chen A, Wu W, Fegley M, Pinnock S, Duffy-Matzner J, Bernier W, Jones W. Pentiptycene-derived fluorescence turn-off polymer chemosensor for copper(II) cation with high selectivity and sensitivity. Polymer. 2017;**9**:118-121. DOI: 10.3390/polym9040118

[79] Wang W, Yang Q, Sun L, Wang H, Zhang C, Fei X, Sun M, Li Y. Preparation of fluorescent nanofibrous film as a sensing material and adsorbent for Cu^{2+} in aqueous solution via copolymerization and electrospinning. Journal of Hazardous Materials. 2011;**194**:185-192. DOI: 10.1016/j.jhazmat.2011.07.083

[80] Li K, Yu R-H, Shi C-M, Tao F-R, Li T-D, Cui Y-Z. Electrospun nanofibrous membrane based on AIE-active compound for detecting picric acid in aqueous solution. Sensors and Actuators B. 2018;**262**:637-645. DOI: 10.1016/j.snb.2018.02.032

[81] Mahmoudifard M, Shoushtari A, Mohsenifar A. Novel approach toward optical sensors based on electrospun nanofibers â quantum dot composits. Textile Industry. 2011;**6**:300-303. http://www.revistaindustriatextila.ro/images/2011/6_2011.pdf

[82] Min M, Wang X, Chen Y, Wang L, Huang H, Shi J. Highly sensitive and selective Cu^{2+} sensor based on electrospun rhodamine dye doped poly(ether sulfones) nanofibers. Sensors and Actuators B. 2013;**188**:360-366. DOI: 10.1016/j.snb.2013.07.017

[83] Quirós J, Amaral A, Pasparakis G, Williams G, Rosal R. Electrospun boronic acid-containing polymer membranes as fluorescent sensors for bacteria detection. Reactive and Functional Polymers. 2017;**121**:23-31. DOI: 10.1016/j.reactfunctpolym.2017.10.007

[84] Shehata N, Samir E, Gaballah S, Hamed A, Elrasheedy A. Embedded ceria nanoparticles in crosslinked PVA electrospun nanofibers as optical sensors for radicals. Sensors. 2016;**16**:1371-1378. DOI: 10.3390/s16091371

[85] Lang C, Fang J, Shao H, Ding X, Lin T. Electrospun nano-nonwoven acoustic sensors. Materials Today. 2017;(4):5306-5311. DOI: 10.1016/j.matpr.2017.05.040

[86] Lin Q, Li Y, Yang M. Investigations on the sensing mechanism of humidity sensors based on electrospun polymer nanofibres. Sensors and Actuators B. 2012;**171-172**:309-314. DOI: 10.1016/j.snb.2012.03.082

[87] Luoh R, Hahn T. Electrospun nanocomposite fiber mats as gas sensors. Composites Science and Technology. 2006;**66**:2436-2441. DOI: 10.1016/j.compscitech.2006.03.012

[88] Wang X, Li Y, Ding B. Electrospun nanofibre-based sensors. In: Ding B, Yu J, editors. Electrospun Nanofibres for Energy and Environmental. Berlin: Springer-Verlag; 2014. pp. 267-297. DOI: 10.1007/978-3-642-54160-5_11

[89] Luo C, Liu N, Zhang H, Liu W, Yue Y, Wang S, Rao S, Yang C, Su J, Jiang X, Gao Y. A new approach for ultrahigh-performance piezoresistive sensor based on wrinkled PPy film with electrospun PVA nanowires as spacer. Nano Energy. 2017;**41**:527-534. DOI: 10.1016/j.nanoen.2017.10.007

[90] Sathiyanathan P, Prabu A, Kim K. Electrospun polyvinylidene fluoride-polyoctafluoropentyl acrylate blend based piezocapacitive pressure sensors. Macromolecular Research. 2016;**24**:670-674. DOI: 10.1007/s13233-016-4098-1

Chapter 4

Fabrication of Ordered and High-Performance Nanostructured Photoelectrocatalysts by Electrochemical Anodization: Influence of Hydrodynamic Conditions

Bianca Lucas-Granados, Rita Sánchez-Tovar, Ramón M. Fernández-Domene and José García-Antón

Additional information is available at the end of the chapter

http://dx.doi.org/10.5772/intechopen.78303

Abstract

Nanostructured semiconductor metal oxides, such as TiO_2, WO_3, Fe_2O_3 or ZnO, are being widely investigated for their use as photoanodes, due to their higher surface areas in contact with the electrolyte, which increases the efficiency of photoelectrochemical processes. Metal oxide nanostructures have been synthesized by a number of different techniques. Anodization is one of the simpler methods used to synthesize nanostructured photoanodes, and the morphology and size of nanostructures can be designed by adequately controlling anodization parameters. Besides, these nanostructures are directly bound to the metallic back contact, improving significantly the efficiency of electron collection. It has been observed that hydrodynamic conditions during anodization (using a rotating disk electrode, RDE) greatly influenced the morphology of nanostructures and, therefore, their photoelectrochemical performance. The objective of this chapter is to review the innovative nanostructures with high-aspect ratios that can be fabricated by anodization under different hydrodynamic conditions.

Keywords: anodization, hydrodynamic conditions, nanostructure, photoelectrocatalysis, semiconductor

1. Introduction

Photoelectrochemistry (PEC) is a multidisciplinary area, which involves electrochemistry, materials science, solid-state physics and optics [1], and studies, in general, processes taking place at the interface between an electrode and an electrolyte, under the influence of an illumination source. The most interesting part of PEC is related to the absorption of incident light in a semiconductor valence band. If the incident light energy is high enough (higher than the band-gap of the semiconductor, $h\upsilon \geq E_g$), this process results in electron excitation from the valence band to the conduction band, hence generating photogenerated electron/hole pairs (e^-/h^+). PEC has numerous attractive applications, such as the split of a water molecule in its fundamental constituents (gaseous hydrogen and oxygen), or the removal of persistent organic pollutants in water treatment processes. The schematic representation of PEC is depicted in **Figure 1**.

In that scheme, a common PEC cell with an n-type semiconductor photoanode and a metallic cathode (typically Pt) is presented. Upon illumination, electrons are promoted from the valence band to the conduction band and are subsequently driven toward the metallic back contact through the semiconductor bulk and, from there, to the cathode through the external circuit. Once in the cathode, these electrons can be used to reduce the oxidized redox species present in the electrolyte, such as water to form gaseous hydrogen. Photogenerated holes, which remain in the semiconductor valence band, are transported toward the semiconductor/electrolyte interface, where they can oxidize the reduced redox species present in the electrolyte, such as water to form gaseous oxygen or oxygen compounds, for example, hydroxyl radical, which could, in turn, directly oxidize recalcitrant organic pollutants (R in **Figure 1**) present in the aqueous solution. The movement of photogenerated charge carriers (e^-/h^+ pairs) described above takes place due to the development of a depleted space-charge region inside

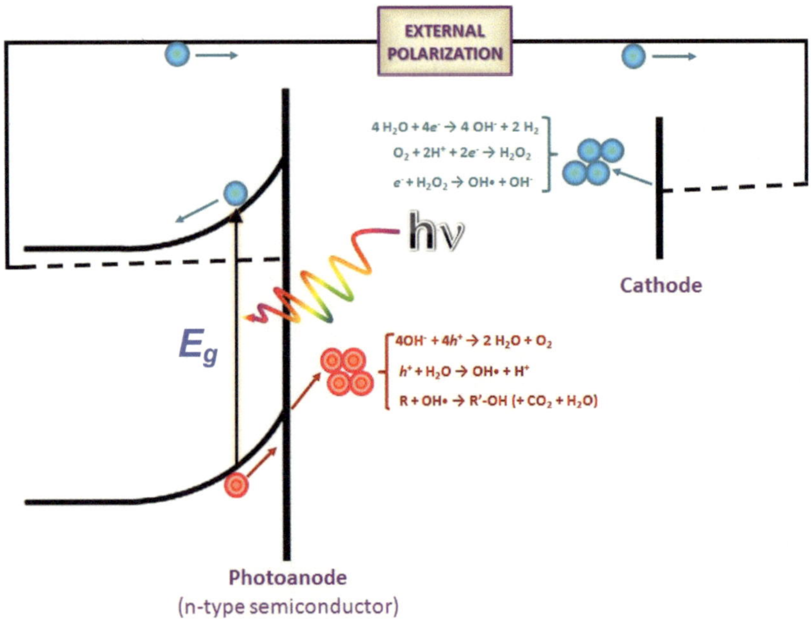

Figure 1. Schematic representation of a photoelectrochemical process.

the semiconductor, where the electric field acts on these charge carriers. To increase the thickness of this space-charge region and to sustain the electric field built up within it, a small external polarization is used in PEC. This polarization significantly improves the transport of electrons from the semiconductor conduction band to the cathode, hence increasing the lifetime of photogenerated charge carriers, minimizing the recombination of these carriers and, therefore, enhancing the PEC process overall efficiency [2–4].

The semiconducting photoanode is the central part of a PEC cell. Consequently, it is essential to develop strategies to fabricate efficient photoanodes, which is a materials science issue. Hence, the development of PEC is closely related to the advances achieved in the design and synthesis of materials used as photoanodes, in which the understanding of their properties and characteristics is fundamental. Among the most important characteristics, an efficient photoanode must satisfy the following [1, 5, 6]:

- **Appropriate bandgap energy (E_g).** The bandgap is the smallest energy difference between the lowest edge of the conduction band and the highest edge of the valence band. A photoanode material should have bandgap energy high enough as to assure that the desired electrochemical reactions can take place (e.g., ~2 eV for water splitting, taking into account electrochemical losses and kinetic aspects [5], but not so high as to severely reduce the amount of usable solar radiation.
- **Appropriate band edge positions.** For hydrogen evolution, the conduction band edge must be positioned above hydrogen evolution, that is, at a more negative potential than that of the normal hydrogen electrode (NHE). For oxygen evolution, the valence band edge must be positioned below the oxygen evolution, that s, at a potential more positive than 1.23 V versus NHE. For generation of hydroxyl radicals, the position of the valence band edge must be located at potentials as high as 2.02 V versus NHE [7].
- **Suitable flat-band potential (E_{FB}).** The flat-band potential is the potential that needs to be applied to the semiconductor to reduce the band bending to zero, that is, at this potential, there is no depleted space charge layer and, therefore, no electric field inside the photoanode. The flat-band potential affects the recombination probability of photogenerated charge carriers: in general, the more negative the value of E_{FB}, the higher the potential drop at the depletion space charge layer and, consequently, the stronger the electrical field within the depleted space charge layer, which is the driving force to separate the photogenerated electron-hole pairs [5, 6, 8].
- **Good electrical conductivity.** Once electrons have been promoted from the valence band to the conduction band of the semiconductor photoanode due to light irradiation, a fast charge separation is crucial for efficient photoelectrochemical processes. For this, the transport of electrons and holes in the material must be good. The electrical conductivity of the photoanode material can be modified by modifying its defect disorder. The density of defects inside the semiconductor (i.e. vacancies, interstitial elements, etc.) must be high enough to enhance electrical conductivity but not so high to negatively affect the efficiency by favoring the recombination of charge carriers.

- **Good chemical, electrochemical and photoelectrochemical stability.** Photoanodes must be stable under prolonged exposure to the electrolyte under the operation conditions (applied potential and illumination).

In recent years, increasing attention is being paid to the design and synthesis of nanostructured metal oxides due to their exceptional properties. Photocurrent density obtained in PEC systems has two main contributions, one coming from the electrocatalytic process and other arising from the photon absorption process. The electrocatalytic process consumes the photogenerated charge carriers and it can become the rate-determining step of the PEC process (the charge transfer at the semiconductor/electrolyte interface) [4]. Current densities in electrochemistry are directly influenced by the real surface active area, since increasing surface area enhances the number of electroactive sites; therefore, nanostructuring the photoelectrode can improve its electrocatalytic performance. Moreover, in nanostructured photoelectrodes, the diffusion path of photogenerated holes toward the semiconductor/electrolyte interface is significantly shorter, which reduces electron-hole recombination. On the other hand, by roughening the semiconductor surface, direct light reflection is reduced due to the scattering of reflected photons, enhancing thus the chance of light absorption by the semiconductor and the possibility of further photoexcitation. Consequently, creating nanostructured photoanodes has immediate benefits on both their electrocatalytic activity and photoactivity [4].

During the last decades, several n-type semiconductor metal oxides, such as TiO_2, WO_3, Fe_2O_3 and ZnO, have been investigated as potential candidates for photoanode materials in PEC cells due to their different characteristics.

Titanium oxide (TiO_2) is a wide bandgap semiconductor ($E_g \approx 3.2$ eV in the anatase crystalline form [3, 4, 9], equivalent to a maximum wavelength of ~390 nm) with an exclusive set of properties, such as high chemical stability, resistance to photocorrosion and favorable band-edge position relative to the redox potentials for the decomposition of water, allowing the use of light for effective direct water splitting [5, 10, 11]. However, due to its high E_g, TiO_2 absorb photons only in the UV range, which represents a very small part of the solar spectrum.

Tungsten oxide (WO_3) is another attractive photoanode material owing to its high electron mobility and moderate hole diffusion length (≈150 nm). Besides, WO_3 can also absorb a part of the visible rays of the solar spectrum (its $E_g \approx 2.6$ eV, which corresponds approximately to a wavelength of $\lambda = 480$ nm) [3, 12, 13]. Moreover, in contrast with almost the rest of semiconductor oxides used as photoelectrocatalysts, WO_3 can safely operate in acidic environments, which is very useful in the treatment of low-pH wastewater [11].

Hematite (α-Fe_2O_3) is the most thermodynamically and stable iron oxide form under ambient conditions and it is a non-toxic, chemically stable and low cost material. Furthermore, it has a bandgap of $E_g \approx 2.1$ eV which indicates its suitability for visible light absorption up to ~590 nm, that is, hematite can collect up to ~40% of solar spectrum energy [4, 11]. Apart from its properties, hematite has some drawbacks such as low carrier mobility and short holes diffusion lengths, which can be overcome by nanostructuring the material.

Zinc oxide (ZnO) is also a wide bandgap semiconductor (E_g = 3.37 eV; λ = 369 nm) with similar characteristics as TiO_2 but with the advantage that ZnO possesses a higher electron mobility (from 10 to 100 times higher in comparison with TiO_2), that is, a large conductivity which makes ZnO a very suitable and promising substitute of TiO_2 [14, 15]. In spite of this, photogenerated electron-hole pairs tend to recombine fast in ZnO, so the use of hybrid semiconductors systems between ZnO and ZnS has been proposed to overcome this inconvenience [16, 17].

In order to prepare nanostructures of the oxides described above, various methodologies have been reported, such as sol-gel processes, hydrothermal and solvothermal methods, deposition processes or anodization. Of these, anodization is a fast and simple method to synthesize metal oxide nanostructures [12, 18–22]. With anodization, surface morphology can be designed by adequately controlling several parameters, such as the anodization potential, duration, electrolyte composition, temperature, and so on. Hence, the morphology and dimensions of nanostructures fabricated by anodization, as well as their PEC behavior, can be controlled by adjusting the aforementioned parameters. Moreover, nanostructures formed by anodization can be grown directly on the substrate (the back metal collector), thus avoiding compaction or sintering of nanostructures. This fact significantly reduces the contact resistance between the nanostructure and the metallic substrate, providing the basis for efficient charge collection. Therefore, anodization offers a robust and economical way to fabricate nanostructured metallic oxides.

Among the parameters that can be controlled to design and fabricate high-performance oxide nanostructures by anodization, hydrodynamic conditions during the synthesis process can be of great importance. Indeed, depending on the used material and the operation conditions, the mechanisms of nanostructures formation by anodization may involve one or more stages where mass transfer becomes the rate-determining step [8, 23]. Therefore, hydrodynamic

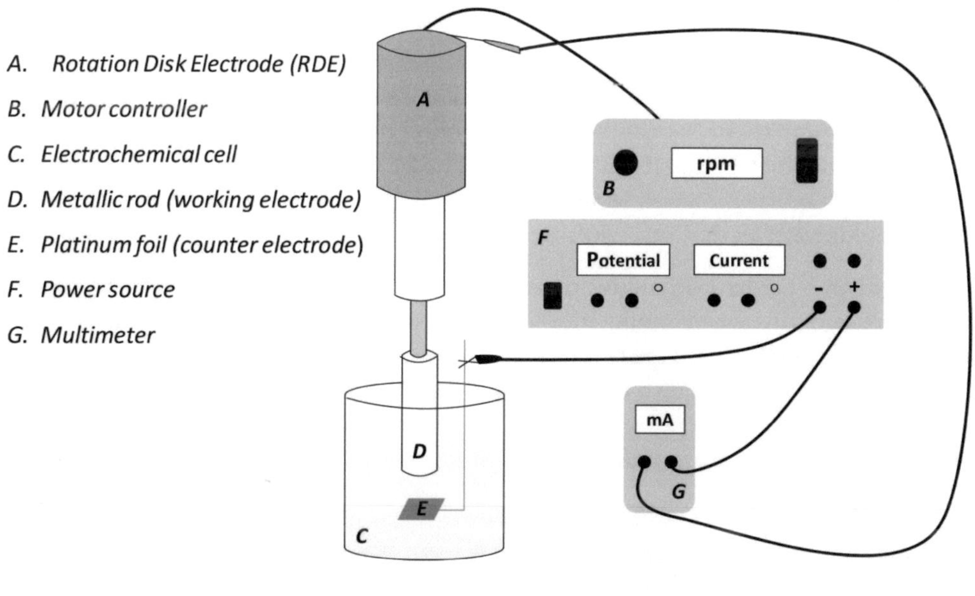

Figure 2. Experimental setup used during all the anodization processes, showing the electrode configuration.

conditions can play a central role in the synthesis of nanostructures by anodization. This chapter presents a review of the innovative high-performance nanostructures that have been fabricated by anodization under different hydrodynamic conditions. Although various fabrication conditions have been used, depending on the material, in all cases, hydrodynamic conditions have been controlled by connecting rods of the different metals to a rotating disk electrode (RDE) to assure reproducible mechanical convection of the electrolyte near the electrode surface. **Figure 2** shows an illustration of the electrode configuration used during all the anodization processes. The specific synthesis conditions for the different materials are detailed in the following sections.

2. TiO_2

Influence of hydrodynamic conditions for nanostructures formed from titanium (Ti) anodization is presented for samples synthesized in two different electrolytes and anodization conditions: ethylene glycol +1 M H_2O containing 0.05 M NH_4F and glycerol/water (60:40 vol.%) containing 0.27 M NH_4F. When anodization was performed in ethylene glycol based electrolytes 55 V for 30 min were directly applied to anodize the Ti. Nevertheless, in glycerol based electrolytes the potential was increased from 0 to 30 V at a rate of 200 mV s^{-1}, applying subsequently the potential of 30 V for 3 h. The active anode area exposed to the electrolyte was 0.5 cm^2.

2.1. Morphological characterization

Since the morphology of the nanostructures is highly dependent on the electrolyte used for anodization, the discussion of the influence of Reynolds number on the morphological properties of the TiO_2 nanostructures will be discussed for each particular anodization media. A field emission scanning electron microscope (FESEM) was used to characterize the morphology of the nanostructures.

Nanostructures anodized in ethylene glycol based electrolytes present an initiation layer which blocks the mouth of the nanotubes, preventing part of the solar radiation from being absorbed at the photoelectrode (**Figure 3a**). It was observed that rotating the electrode during anodization, part of the initiation layer was removed, and this elimination was more important as Reynolds number (Re) was increased (**Figure 3b**).

On the other hand, the hydrodynamic conditions completely changed the morphology of the formed nanostructures when anodization was carried out in glycerol based electrolytes, that is from nanotubes (at Re = 0, **Figure 4a**) to nanosponges (at Re > 0, **Figure 4b**). Nanosponge morphology is characterized by a connected and highly porous TiO_2 structure.

2.2. Anodization profile: current densities vs. time

The formation and growth of the TiO_2 nanostructures was monitored by recording the current density-time behavior (**Figure 3c** and **4c**). The current density profile follows a similar trend regardless the electrolyte used for anodization.

http://dx.doi.org/10.5772/intechopen.78303

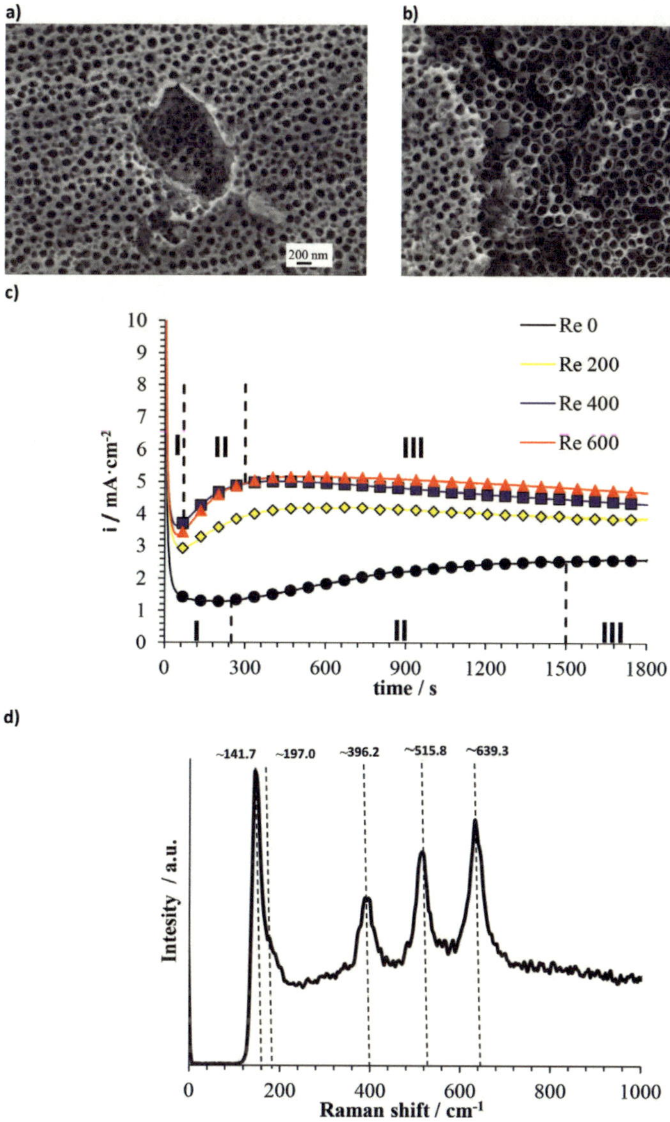

Figure 3. FESEM images of the nanostructures anodized in ethylene glycol-based electrolytes at a Re 600 (a and b). Current density versus time profiles of the nanostructures during anodization in ethylene glycol based electrolytes at the different hydrodynamic conditions (Re from 0 to 600) (c). Raman spectrum of a nanostructure anodized at re = 600 in ethylene glycol-based electrolytes and annealed in air at 450°C for 1 h (d).

In the first stage (I), current density decreases with time due to a compact TiO_2 oxide layer formation [9, 24]. Then, in the second stage (stage II), an increase in current density indicates the onset of TiO_2 dissolution due to the presence of fluoride ions in the electrolyte (stage II), according to the reaction shown in Eq. 1 [25–27].

$$TiO_2 + 6F^- + 4H^+ \rightarrow [TiF_6]^{2-} + 2H_2O \quad (1)$$

After that, in the third stage (stage III), current density remains almost constant due to the formation and growth of regular nanotubes or nanosponges [9, 28].

In general, **Figures 3c** and **4c** show that current densities increase with increasing Re in all the stages, which indicates that hydrodynamic conditions enhance the electrochemical processes taking place during anodization.

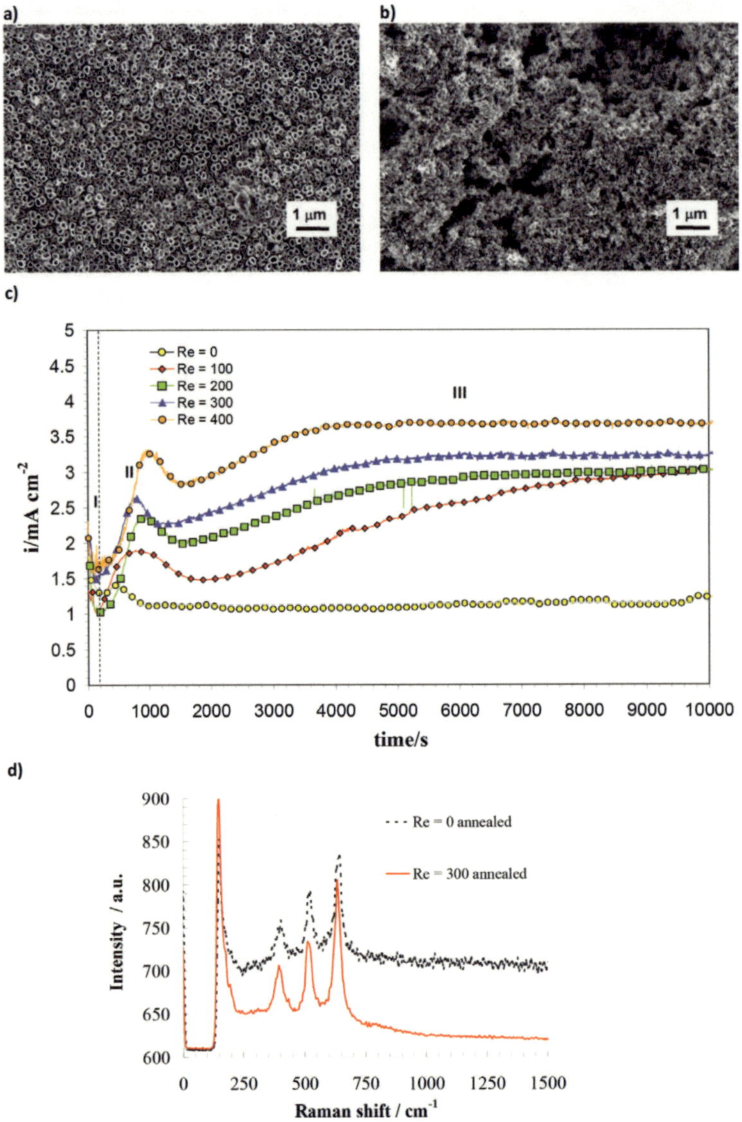

Figure 4. FESEM images of the nanostructures anodized in glycerol based electrolytes at a re = 0 (a) and re = 300 (b). Current density vs. time profiles of the nanostructures during anodization in ethylene glycol based electrolytes at the different hydrodynamic conditions (re from 0 to 400) (c). Raman spectra of the nanostructures anodized in glycerol based electrolytes, at re = 0 and re = 300 and annealed in air at 450°C for 1 h (d).

http://dx.doi.org/10.5772/intechopen.78303

2.3. Crystalline structure characterization

In order to obtain a crystalline structure, samples were annealed in air at 450°C for 1 h. After the annealing treatment, all the samples possessed the Raman peaks of the TiO_2 anatase phase shown in **Figures 3d** and **4d** (peaks at roughly 141.7, 197.0, 396.2, 515.1 and 639.3 cm^{-1} [28–30]).

2.4. Electrochemical and photoelectrochemical characterization

Electrochemical impedance spectroscopy (EIS) measurements at open circuit potential (OCP) and Mott-Schottky plots were performed in 0.1 M Na_2SO_4 in order to evaluate the electrochemical properties of the samples. **Figure 5** shows the Nyquist (**Figure 5a**), Bode-phase (**Figure 5b**) and Bode-modulus (**Figure 5c**) plots for the nanostructures anodized in ethylene glycol-based electrolytes.

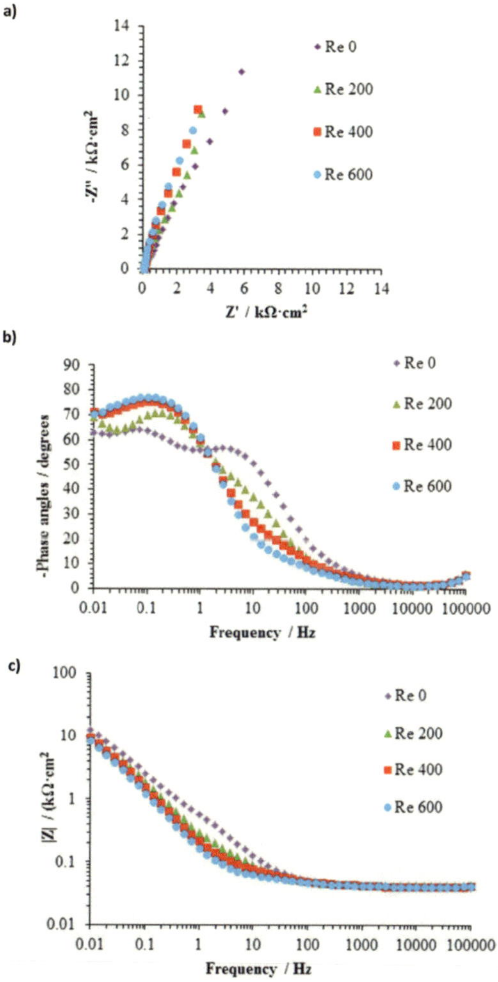

Figure 5. Experimental Nyquist (a), bode-phase (b) and bode-modulus (c) plots for the nanostructures anodized in ethylene glycol based electrolytes at Reynolds numbers = 0, 200, 400 and 600. Experiments performed in 0.1 M Na_2SO_4.

Figure 6 shows the electrical equivalent circuit used to fit EIS data, where R_s is the electrolyte resistance, and the two groups of resistances and constant phase elements (R-CPE) correspond to the nanotubular layer (R_1-CPE_1) and the compact TiO_2 underlayer (R_2-CPE_2) [8, 31, 32].

The different parameters obtained from the fitting of the EIS data in the electrical equivalent circuit are shown in **Table 1**. From **Table 1**, it is important to highlight that the resistance of the compact layer is always higher than the one obtained for the nanotubes ($R_2 > R_1$), due to the higher conductivity of the nanostructures. On the other hand, R_1 decreases as Reynolds number increases, that is, the conductivity of the nanotubes increases at higher hydrodynamic conditions.

In order to observe the electrochemical behavior of nanostructures formed in glycerol-based electrolytes, **Figure 7** shows the Mott-Schottky (MS) plots at a frequency of 10 kHz (to eliminate the capacitance dependence on frequency) for nanotubes (Re = 0) and nanosponges (Re > 0).

In **Figure 7**, positive slopes of the MS plots are characteristic of *n*-type semiconductors. The dominant TiO_2 defects are oxygen vacancies due to their lower formation energy compared with Ti^{3+} interstitials [32–38]. **Table 2** shows the donor densities (N_D) and the flat band potential (E_{FB}) values at different Re.

The donor density can be determined from the positive slopes of the straight lines in the MS plots using the Mott-Schottky equation for an *n*-type semiconductor:

$$\frac{1}{C^2} = \frac{2}{\varepsilon_r \varepsilon_0 e N_D}\left(E - E_{FB} - \frac{kT}{e}\right) \quad (2)$$

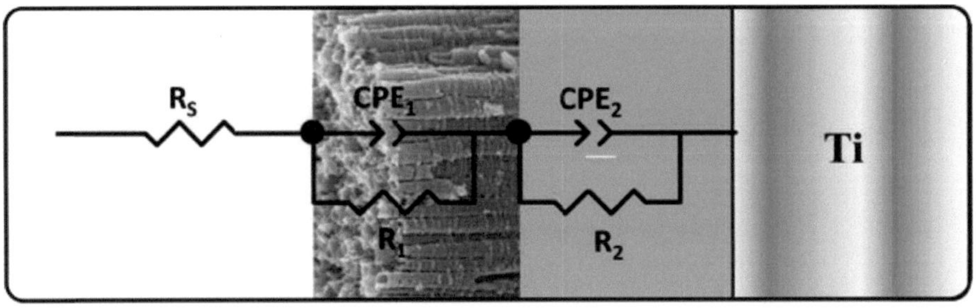

Figure 6. Electrical equivalent circuit used to simulate the experimental data obtained from EIS experiments shown in **Figure 5**.

Re	R_S/Ω cm²	$C_1/\mu F$ cm-2	α_1	$R_1/k\Omega$ cm2	$C_2/\mu F$ cm⁻²	α_2	$R_2/k\Omega$ cm²	χ^2 (x10⁻³)
0	38 ± 4	106 ± 14	0.68 ± 0.03	5.7 ± 1.8	1690 ± 78	0.98 ± 0.02	78.3 ± 21	5.3
200	39 ± 3	279 ± 11	0.63 ± 0.04	5.1 ± 1.3	1903 ± 84	0.95 ± 0.01	39.4 ± 23	9.6
400	42 ± 5	475 ± 17	0.56 ± 0.08	4.4 ± 1.5	1942 ± 101	0.95 ± 0.03	24.8 ± 12	8.5
600	34 ± 7	132 ± 18	0.74 ± 0.07	2.5 ± 1.0	2399 ± 97	0.97 ± 0.04	38.4 ± 11	8.0

Table 1. Experimental values obtained from the equivalent circuit fitting for the nanostructures anodized in ethylene glycol-based electrolytes at the different Re = 0, 200, 400 and 600.

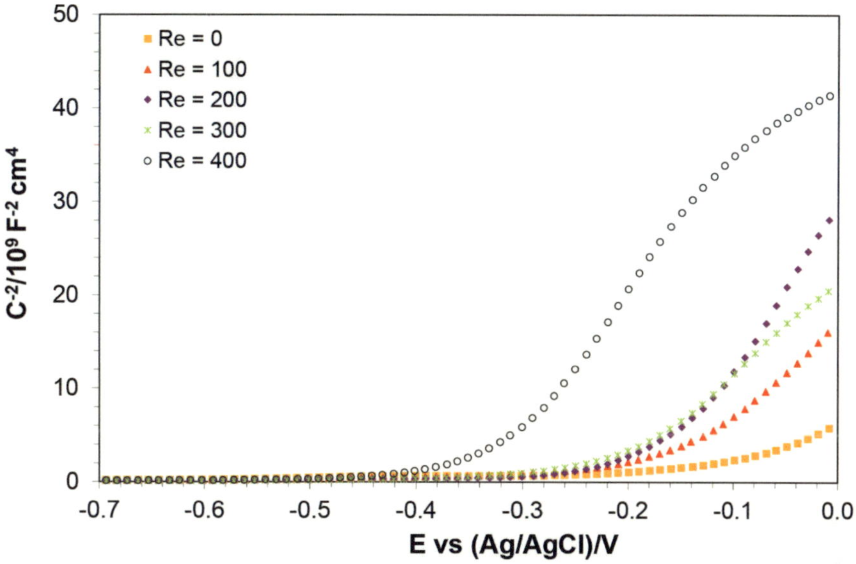

Figure 7. Mott-Schottky plots of the different nanostructures (nanotubes at Re = 0 and nanosponges at Re > 0) anodized in glycerol based electrolytes. Experiments performed in 0.1 M Na_2SO_4.

Re	N_D ($\times 10^{19}$ cm^{-3})	E_{FB}/V vs (Ag/AgCl)
0	3.6 ± 0.6	-0.15 ± 0.07
100	1.4 ± 0.3	-0.17 ± 0.04
200	1.1 ± 0.1	-0.17 ± 0.02
300	1.1 ± 0.2	-0.21 ± 0.01
400	0.9 ± 0.1	-0.33 ± 0.03

Table 2. Values of N_D and E_{FB} for the different nanostructures synthetized in glycerol-based electrolytes under static (Re = 0) and under hydrodynamic conditions (Re = 100, 200, 300 and 400).

where ε is the dielectric constant of the TiO_2 layers (100 for TiO_2 nanostructures [39]), ε_0 is the vacuum permittivity ($8.85 \cdot 10^{-14}$ F/cm), e is the electron charge ($1.60 \cdot 10^{-19}$ C) and σ is the positive slope of each straight line in the MS plots.

Table 2 shows that N_D decreases with increasing Re, that is, N_D is lower for the nanosponges than for the nanotubes. In this case, the increase in N_D leads to a decrease in the depletion layer thickness, resulting in an increase of recombination losses [6, 40, 41]. In fact, oxygen vacancies in TiO_2 act as recombination centers for electron and holes, playing a critical role in the trapping process [42–44].

Table 2 also shows the flat band potentials for the different nanostructures obtained in glycerol-based electrolytes. These values were determined from the intercept of the straight line in MS plots with the potential axis. To reduce the recombination probability, flat band potentials should be high and negative [6, 42]. It can be observed in **Table 2** that flat band potentials are more negative with increasing Re, that is, nanosponges formed under hydrodynamic conditions possess more negative flat band potentials than nanotubes formed under static conditions. Thus, the charge recombination probability is lower in the case of the nanosponges.

Figure 8 shows the photoelectrochemical water splitting performance under simulated sunlight AM 1.5 conditions in 1 M KOH for the different nanostructures anodized in ethylene glycol (**Figure 8a**) and glycerol (**Figure 8b**)-based electrolytes.

It can be observed in terms of the photocurrent transient vs. potential curves that the nanostructures obtained under hydrodynamic conditions present a higher performance in comparison with the nanostructures anodized at the same potential, time and electrolyte but at

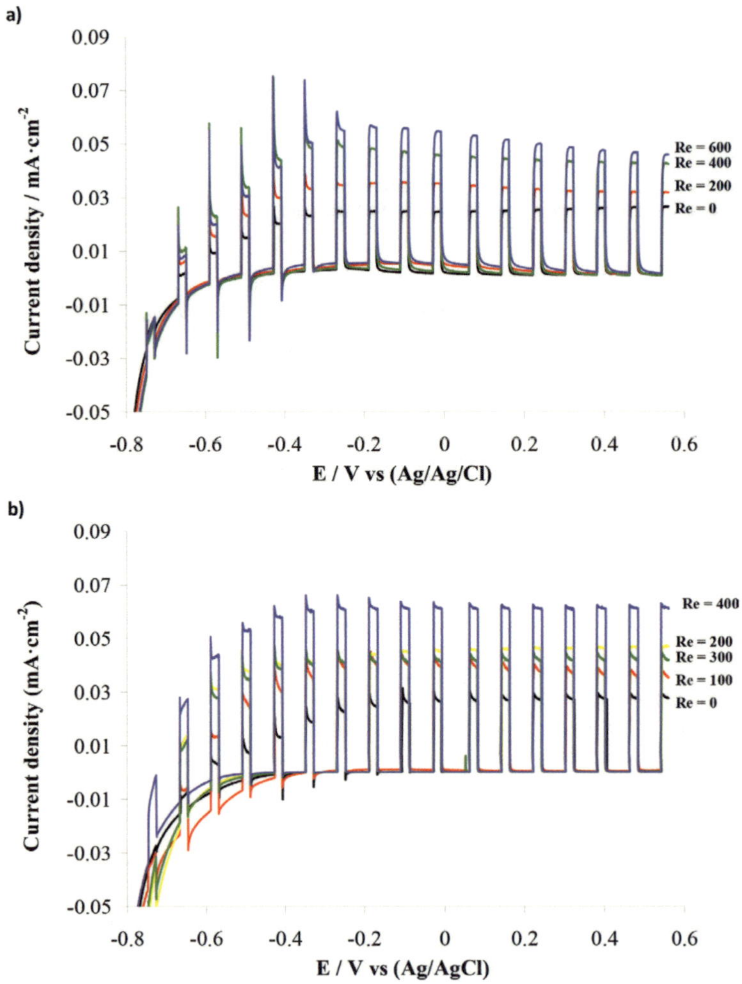

Figure 8. Current density versus applied potential profiles under dark and illumination conditions for the different nanostructures anodized under static (Re = 0) and at different hydrodynamic conditions (Re > 0) in ethylene glycol (a) and glycerol (b)-based electrolytes. Experiments were performed in 1 M KOH.

Re = 0. These results are in agreement with the morphology of the samples (high surface area for the nanostructures anodized under hydrodynamic conditions), EIS measurements (low resistance values, i.e., high conductivity for nanostructures synthetized at Re > 0) and MS plots (appropriate donor density values and more negative flat band potentials).

Therefore, on the one hand, hydrodynamic conditions are beneficial in ethylene glycol-based electrolytes because of the removal of the initiation layer which offers a high surface area for photoelectrochemical water splitting. On the other hand, nanosponges formed in glycerol based electrolytes under dynamic conditions enhance the rate of the photoelectrochemical reaction over the photoelectrode due to their surface-active sites.

3. WO_3

Different WO_3 nanostructures synthesized by anodization have been presented in the literature in the past few years. Several morphologies have been obtained, such as nanopores [19, 45–52], nanoplatelets [53–55], nanoparticles [56], flower-shape nanostructures [57], and so on. Recently, we have presented novel WO_3 nanostructures fabricated by anodization under different controlled hydrodynamic conditions and, in general, photoresponses significantly higher than those for nanostructures synthesized under static conditions were obtained.

These nanostructures were produced by anodization of tungsten rods in sulfuric acid media in the presence of small amounts of complexing agents (NaF and H_2O_2). For example, tree-like globular clusters of WO_3 nanoplatelets were obtained in a 1.5 M H_2SO_4 + 0.1 M NaF electrolyte and at different rotation velocities of the RDE. **Figure 9** shows the FE-SEM images of the samples anodized in that electrolyte at different rotation velocities (0 and 375 rpm) [23, 58]. It can be observed that in both cases, the nanostructures formed on the tungsten surface presented nanoplatelet morphology, as other authors reported when anodizing in similar acid electrolytes [53–55, 57, 58]. Nevertheless, the morphology and dimensions of these nanoplatelets varied noticeably with rotation speed. At 0 rpm (**Figure 9a**), nanoplatelets grew quite orthogonal to the electrode surface, in a more or less ordered way. At 375 rpm (**Figure 9b**), nanoplatelets formed globular clusters that grew in a tree-like manner, forming a multilevel WO_3 layer resembling a forest seen from above. It is perceptible that the new nanostructure obtained under controlled hydrodynamic conditions exposed more surface area to the electrolyte than the one synthesized at 0 rpm.

In the presence of other well-known complexing agent of tungsten—hydrogen peroxide—and under controlled hydrodynamic conditions, a different nanostructure was obtained [59]. In this case, anodization of tungsten in a 1.5 M H_2SO_4 + 0.05 M H_2O_2 and at 375 rpm (the same rotation speed as in the example shown above for the NaF electrolyte), resulted in very small nanoplatelets or nanosheets forming a kind of spongy layer (**Figure 10**).

The three nanostructures shown in **Figures 9** and **10** shared a common formation mechanism, that is, a dissolution/precipitation mechanism, although it was greatly influenced by hydrodynamic conditions. In order to study the influence of rotation speed of the electrode on the

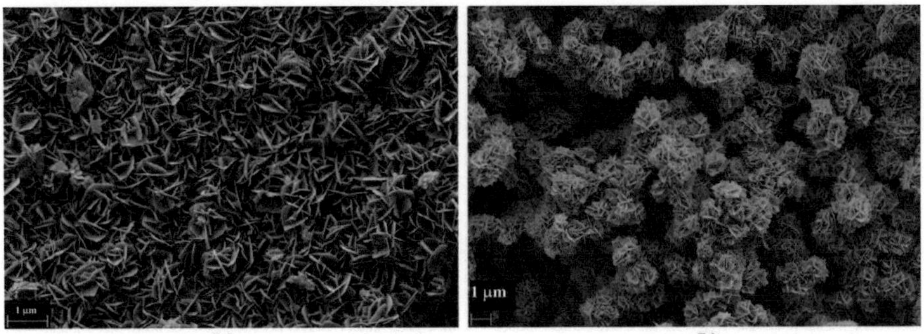

(a) (b)

Figure 9. FE-SEM images of the samples anodized in the 1.5 M H_2SO_4 + 0.1 M NaF electrolyte at 0 rpm (a) and 375 rpm (b).

Figure 10. FE-SEM images of the samples anodized in the 1.5 M H_2SO_4 + 0.05 M H_2O_2 electrolyte at 375 rpm and at different magnifications.

formation and growth of WO_3 nanoplatelets, current density transients were recorded during the anodization processes. **Figure 11** shows, by way of illustration, current density transients recorded for the WO_3 nanostructures anodized in the 1.5 M H_2SO_4 + 0.1 M NaF electrolyte, under static (0 rpm) and hydrodynamic conditions (375 rpm).

During the first seconds of anodization, a notable decrease in current density can be observed for both cases. This decrease has been associated with the formation of a compact WO_3 layer on the electrode surface [19, 23, 58, 60]. Subsequently, current density started increasing, indicating the dissolution of the WO_3 layer due to the effect of H^+ and F^- [23, 58, 61]. This dissolution process resulted in the release of soluble cationic species, such as WO_2^{2+} (or $[W(OH)_4(H_2O)_4]^{2+}$ in its hydrated form) or $[WF_n]^{(6-n)}$, from the electrode surface. Afterward, current density gradually decreased again, eventually reaching a steady-state value. This decrease and the later stabilization of current density can be explained by the precipitation of soluble species upon reaching supersaturation conditions near the electrode surface. In fact, as the dissolution of the WO_3 layer proceeded, increasing amounts of tungsten cationic species were formed and they reacted to produce polycondensed tungstates [62]. These species finally precipitated on the electrode surface in the form of insoluble and highly polymeric tungstic acids ($WO_3 \bullet H_2O$ and $WO_3 \bullet 2H_2O$) due to strongly acidic pH of the electrolyte

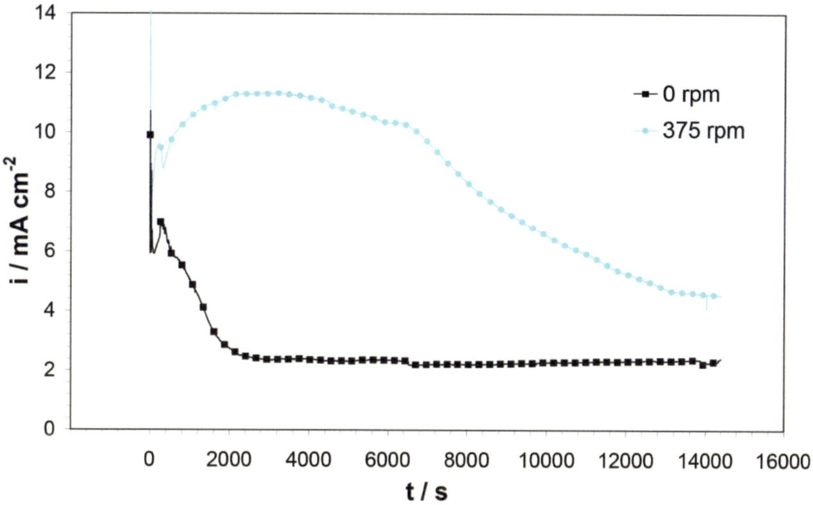

Figure 11. Current density transients recorded during anodization for the WO_3 nanostructures fabricated in the 1.5 M H_2SO_4 + 0.1 M NaF electrolyte, under static (0 rpm) and hydrodynamic conditions (375 rpm).

[59, 62]. The particular morphologies of nanoplatelets were a consequence of the interaction between soluble tungsten species and fluoride anions during the precipitation process [59], being hydrodynamic conditions especially important [23].

Under static conditions (0 rpm), current densities recorded during the whole anodization process were notably lower than under hydrodynamic conditions. Moreover, steady-state values were attained much earlier than with rotation, indicating that the WO_3 nanoplatelet layer completely covered the electrode surface after ~2000 s from the beginning of the anodization process. These facts imply that the electrode rotation permitted the ions in the electrolyte (F^- and H^+) to transport to the electrode more rapidly, resulting in higher current densities being measured at the working electrode due to the enhancement of the WO_3 compact layer dissolution. Consequently, an increase in the rotation speed increased the release of soluble species from the electrode surface, which would result in a higher precipitation rate of hydrated WO_3 in the form of nanoplatelets.

Photoelectrochemical characterization of the nanostructures was performed under simulated sunlight illumination AM 1.5 (100 mW cm^{-2}) imposing an external polarization of 1 V (vs Ag/AgCl). Photocurrent density-potential curves for the three different WO_3 nanostructures presented in **Figures 9** and **10** are shown in **Figure 12**. It can be observed that nanoplatelets anodized under static conditions (0 rpm) provided the lowest photocurrent density (~0.41 mA cm^{-2} at 1 V). Nanostructures anodized under hydrodynamic conditions (in the different electrolytes) showed a large enhancement of their photoelectrocatalytic activity (~1.51 mA cm^{-2} at 1 V and ~1.86 mA cm^{-2} at 1 V for nanostructures anodized in the presence of NaF and H_2O_2, respectively). This clear improvement is directly related to the increase of the electrochemically active surface area in nanostructures synthesized under hydrodynamic conditions. As explained earlier, nanoplatelets formed in the presence of NaF and with rotation grew in a tree-like manner forming globular clusters, which made this new nanoplatelet configuration

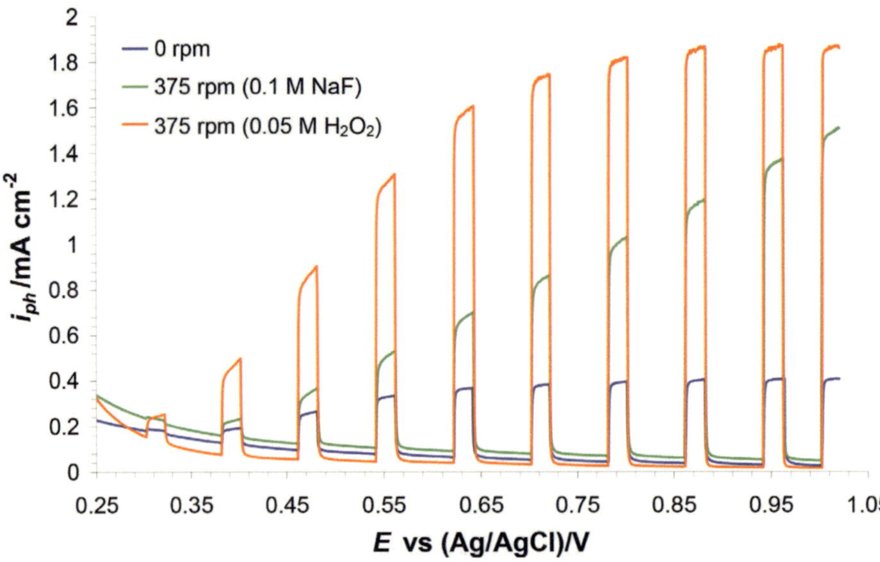

Figure 12. Photocurrent density-potential curves for the three different WO_3 nanostructures presented in **Figures 9** and **10**.

expose much more active area than normal nanoplatelets fabricated at 0 rpm. In the presence of 0.05 M H_2O_2, the obtained nanoplatelets were very small and they aggregated forming very thin layers, also resulting in an increase in the surface area with respect to other WO_3 nanoplatelets and in higher photoelectrochemical efficiencies.

4. Fe_2O_3

Iron oxide nanostructures were made by electrochemical anodization in order to evaluate the effect of hydrodynamic conditions on the formed nanostructures. Prior to anodization, iron rods were abraded with SiC papers of 220, 500 and 4000, sonicated in ethanol, rinsed with distilled water and dried in nitrogen stream. Electrochemical anodization was carried out at room temperature with an ethylene glycol-based solution with 0.1 M NH_4F and 3% vol. H_2O for 15 min at 50 V with iron rod as working electrode and a platinum foil as counter electrode [63]. Different rotation speeds: 0, 1000, 2000 and 3000 rpm, corresponding to Reynolds numbers of 0, 165, 325 and 490, respectively, were applied. Once synthesized, samples were annealed in argon atmosphere for 1 h at 500°C at a heating rate of 15°C · min^{-1}, and cooled within the furnace by natural convection [21].

4.1. Electrochemical anodization

Figure 13 shows the different obtained curves during anodization where three stages can be seen for all the cases (scheme in **Figure 13**). In the first stage, an abrupt drop in the current density from 80 mA · cm^{-2} to less than 15 mA·cm^{-2} (in the case of 0 rpm) and less than 25 mA · cm^{-2} (in the case of hydrodynamic conditions), indicated a formation of an insulating compact oxide layer (Fe_2O_3) on the iron substrate according to Eq. (3).

http://dx.doi.org/10.5772/intechopen.78303

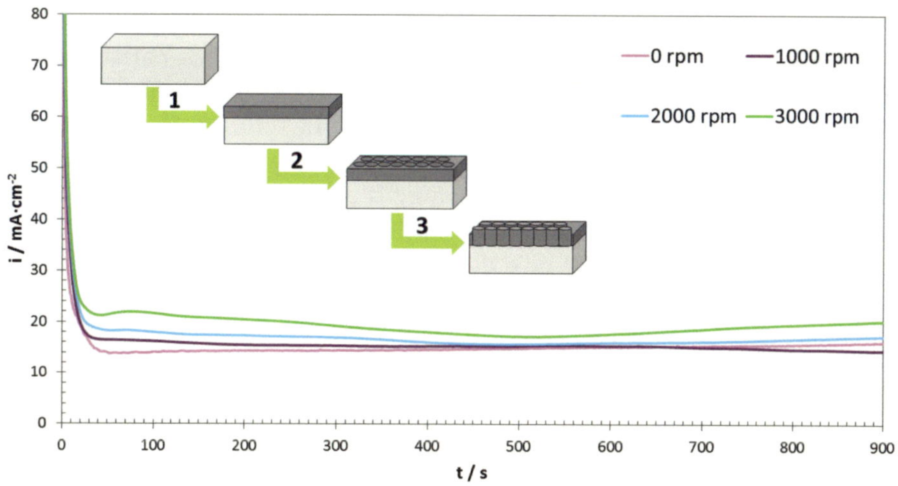

Figure 13. Current density vs. time curves during electrochemical anodization of iron under different electrode rotation speeds. Inset shows a scheme of the process occurring in the electrode during anodization.

In the second stage, a slight increase in current density indicated that tiny pits occur in the surface of the compact layer, followed by the formation of the nanostructures resulting in a decrease in resistance. This process occurs because of the fluoride ions (F^-) and the applied potential which lead to partial dissolution of the compact layer forming the nanoporous structure as Eq. (4) indicates.

$$2Fe + 3H_2O \rightarrow Fe_2O_3 + 3H_2 \quad (3)$$

$$Fe_2O_3 + 12F^- + 6H^+ \rightarrow 2[FeF_6]^{3-} + 3H_2O \quad (4)$$

Finally, in the third stage, further dissolution and cation-cation repulsion occurs and the nanoporous structure leads to a nanotubular one. This occurs until equilibrium between the formation of the oxide layer and its chemical dissolution by F^- is reached. In this way, current density remains almost constant and the formation of nanotubular structure stops [64, 65].

According to **Figure 13**, the formation of the nanostructure under stagnant conditions (0 rpm) leads to lower values of the current density in comparison to the ones synthesized under hydrodynamic conditions. In fact, the higher the rotation speed the higher the current density, achieving the maximum value for the samples anodized at 3000 rpm. This is because steady-state current density is controlled by diffusion processes, hence stirring the iron rods during anodization increases the diffusion and then, current density increases [66, 67].

4.2. Structural characterization

A field emission scanning electron microscope was used in order to evaluate the morphology of the synthesized nanostructures. Moreover, crystalline structure of the samples was evaluated by a confocal Raman microscope with the aim to determine crystalline phases present in the nanostructures.

Figure 14a shows the morphology obtained for the samples under stagnant conditions. In this case, a cracked porous initiation layer appears over the nanotubes, partially covering the entrances of the tubes. This initiation layer results in a decrease in the efficiency of the nanostructures since they might be less accessible to light. On the contrary, **Figure 14b** shows that the initiation layer is etched and the nanotubes are accessible to light irradiation, which leads to higher photoactivity. Nanotubular morphology is advantageous for photoelectrochemical water splitting since it improves electron transport behavior with the tubular orientation. Furthermore, nanotubes avoid short hole diffusion lengths problem since tubes walls make shorter the route from the places where the holes are generated to the surfaces where oxidation reactions are occurring [68].

Figure 14c illustrates the morphology of the samples synthesized at 2000 rpm, and at this rotation speed, the nanotubes collapse and seem to be stacked, so the nanotubes are unseen

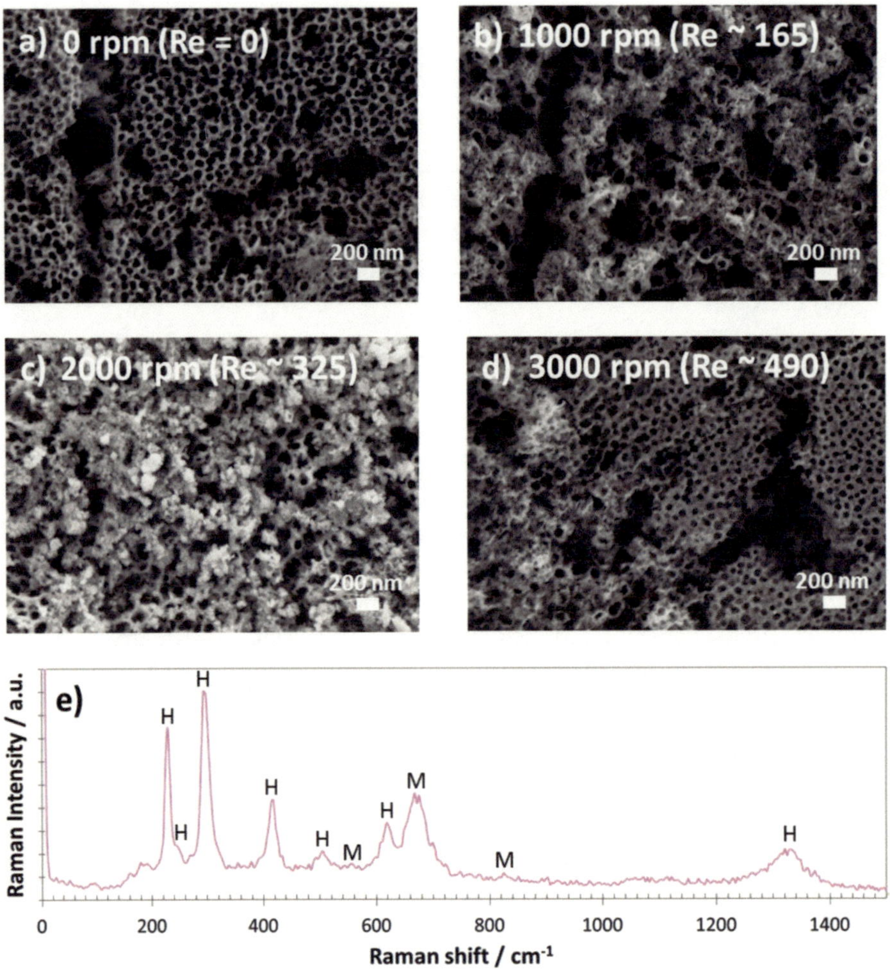

Figure 14. FE-SEM images of the nanostructures synthesized under different electrode rotation speeds: (a) 0 rpm, (b) 1000 rpm, (c) 2000 rpm and (d) 3000 rpm, and (e) Raman spectrum of the nanostructures.

and the top morphology seems to be chaotic. This morphology results in less photoactivity since light irradiation cannot go deep into the nanotubes. On the other hand, when rotation speed is 3000 rpm (see **Figure 14d**), the morphology is a mixture between what occurs at 0 and 2000 rpm, that is, the nanotubes are collapsed and stacked and an initiation layer appears in some parts of the nanostructure covering the tubes [69]. Hence, illumination is not effective in this case, and the photoactivity is affected. This could be due to the fact that at 3000 rpm, some vortex could be formed and then the conditions are non-homogeneous during anodization.

On the other hand, thickness of the nanostructures were measured, and they were comprised between 810 and 870 nm in all cases, which indicated that rotation speed does not affect the thickness of the synthesized nanostructures [69].

Raman spectroscopy was measured with a neon laser 632 nm with ~700 μW. Raman spectra of all the samples were the same, suggesting that rotation speed does not change the crystalline structure of the formed nanostructures. **Figure 14e** shows a Raman spectrum as an example. The Raman shift at which appear the peaks indicates that the nanostructures are mainly composed of hematite with peaks at roughly 229 cm^{-1} (A_{1g}), 249 cm^{-1} (E_g), 295 cm^{-1} (E_g), 414 cm^{-1} (E_g), 500 cm^{-1} (A_{1g}), 615 cm^{-1} (E_g) and 1317 cm^{-1} (2nd order). However, some peaks appearing at approximately 554 cm^{-1}, 672 cm^{-1} and 820 cm^{-1} also indicated the presence of some amount of magnetite [70, 71].

4.3. Photoelectrochemical water splitting tests

For the photocurrent density versus potential measurements the potential was varied from −0.4 to +0.6 $V_{Ag/AgCl}$ at a scan rate of 2 mVs^{-1}.

Figure 15 shows that the highest photocurrent density values were achieved for the nanostructure synthesized at 1000 rpm (reaching 0.130 mA · cm^{-2} at 0.54 $V_{Ag/AgCl}$), which indicated its suitability as photocatalyst for water splitting [69]. This is in agreement with FE-SEM images

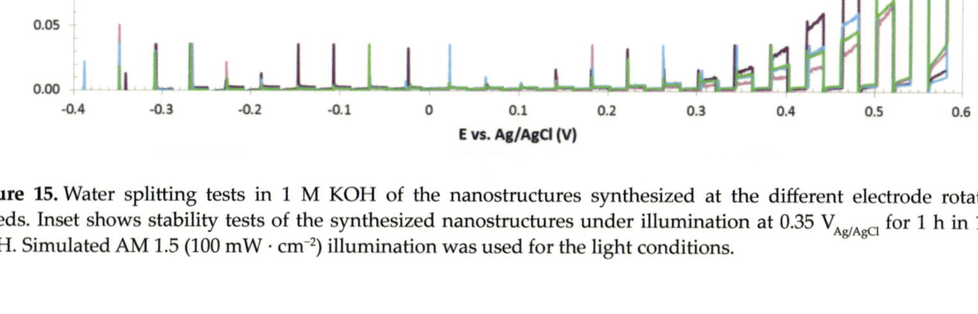

Figure 15. Water splitting tests in 1 M KOH of the nanostructures synthesized at the different electrode rotation speeds. Inset shows stability tests of the synthesized nanostructures under illumination at 0.35 $V_{Ag/AgCl}$ for 1 h in 1 M KOH. Simulated AM 1.5 (100 mW · cm^{-2}) illumination was used for the light conditions.

since they revealed that the morphology of the nanostructures synthesized at 1000 rpm was the most adequate for being illuminated in photoelectrochemical measurements.

Furthermore, inset of **Figure 15** indicates that all synthesized nanostructures were stable against photocorrosion at the indicated conditions.

4.4. Mott-Schottky analysis

For the Mott-Schottky analysis, the potential was swept from the OCP value (~ −0.3 V) in the negative direction at 28 mV s^{-1} with an amplitude signal of 0.01 V at a frequency value of 5 kHz.

Figure 16 shows MS plots for all the synthesized nanostructures under dark (A) and light (B) conditions, indicating a positive slope corresponding to an n-type semiconductor. Note that the higher the slopes of the linear region in MS plots, the lower the electron donor density according to Mott-Schottky equation used for an n-type semiconductor (Eq. (2)), with an assumed ε_r of 80 for the nanostructures [72, 73].

Table 3 shows the results for the different nanostructures and all the values are in the order of 10^{19} cm^{-3}, regardless the rotation speed during anodization. However, the samples anodized at 0 and 3000 rpm achieved donor density values too high which is detrimental for the

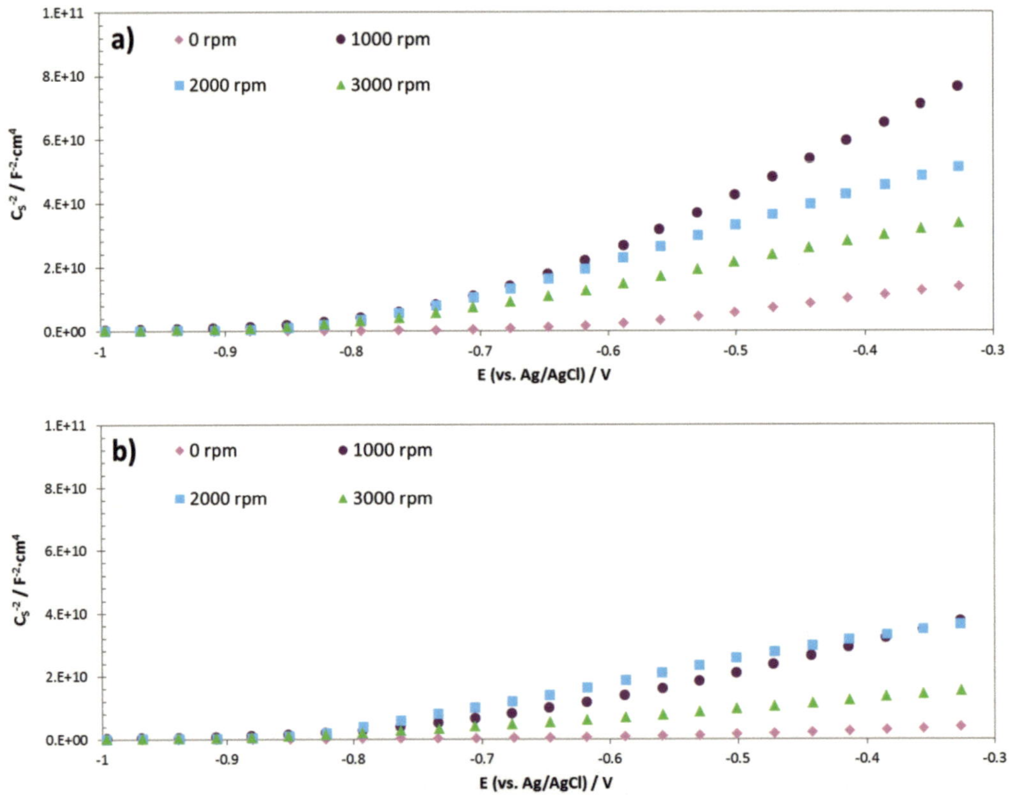

Figure 16. MS analysis of the nanostructures synthesized at the different electrode rotation speeds under dark (a) and 1.5 AM (100 mW · cm^{-2}) illumination (b) conditions in 1 M KOH.

Rotation speed/rpm	Conditions	N_D (10^{19})/cm^{-3}	$E_{FB}/V_{Ag/AgCl}$
0 rpm	Dark	4.0 ± 1.0	−0.64 ± 0.05
	Light	14.3 ± 3.5	−0.63 ± 0.05
1000 rpm	Dark	1.0 ± 0.2	−0.76 ± 0.04
	Light	2.2 ± 0.7	−0.78 ± 0.03
2000 rpm	Dark	1.6 ± 0.6	−0.80 ± 0.08
	Light	2.4 ± 0.3	−0.85 ± 0.09
3000 rpm	Dark	2.5 ± 1.0	−0.80 ± 0.03
	Light	5.8 ± 0.7	−0.82 ± 0.03

Table 3. Values of donor density (N_D) and flat band potential (E_{FB}) for the samples synthesized at the different electrode rotation speeds.

photoelectrochemical water splitting since the defects could act as carrier traps (trapping the electrons or holes) [42]. Hence, the efficiency of the water splitting is lower as shown in **Figure 16.** On the contrast, lower donor density values, as in the case of the nanostructure synthesized at 1000 rpm, achieved better photoelectrochemical results (see **Figure 16**) since it avoids recombination processes. This indicates that donor density is a parameter that considerably affects the efficiency of the water splitting.

On the other hand, flat band potential (E_{FB}) is related to the potential drop at the depletion space charge layer ($\Delta\phi_{SC}$) and to the applied external potential (E) according to Eq. (5). Then, the higher and more negative flat band potential, the lower recombination rate [6].

$$\Delta\phi_{SC} = E - E_{FB} \tag{5}$$

Table 3 shows that the flat band potential does not vary for the different nanostructures, which in fact demonstrates that electrode rotation speed during anodization does not affect flat band potential of the synthesized nanostructures.

5. ZnO/ZnS heterostructures

The influence of hydrodynamic conditions was studied for ZnO/ZnS heterostructures obtained from anodization of zinc (0.5 cm^2 area exposed to the electrolyte) in glycerol/water (60:40 vol%) media with 0.025 M NH_4F and 0.2 M Na_2S. Nanostructures were synthesized at different voltages: 20, 30 and 40 V. The advantage of this type of heterostructure is the good photocatalytic behavior of ZnS, which in combination with ZnO could reduce the recombination rate of the charge carriers due to their separated band gaps.

5.1. Morphological characterization

As an example of the nanostructures of ZnO/ZnS obtained from anodization of zinc, **Figure 17** shows the FESEM images of top and cross-sectional views of the samples anodized at different voltages and hydrodynamic conditions. It has been demonstrated [22, 74] that anodization of zinc in aqueous electrolytes containing both sulfide and fluoride generates a top compact layer and an underneath nanostructure. The former mainly consists of ZnS and the latter of ZnO. That is, the ZnO/ZnS heterostructures grow under the compact ZnS layer. In this way, **Figure 17a** zinc shows the top surface of the nanostructure obtained at 20 V and under static conditions, there, it can be clearly observed the rough ZnS top layer surface. Under stagnant conditions (0 rpm) the anodization of zinc is not homogeneous, since bubbles are formed at the anode surface, and consequently, the formation of precipitates takes place. On the other hand, **Figure 17b** and **c** shows that, when anodization is performed stirring the zinc road (under hydrodynamic conditions), a smooth ZnS top layer is formed. If we compare the heterostructures synthetized at 1000 and at 3000 rpm (**Figure 17b** and **c**, both at 20 V), there is a change in the morphology of the ZnO/ZnS samples from a porous sponge type (**Figure 17b**)

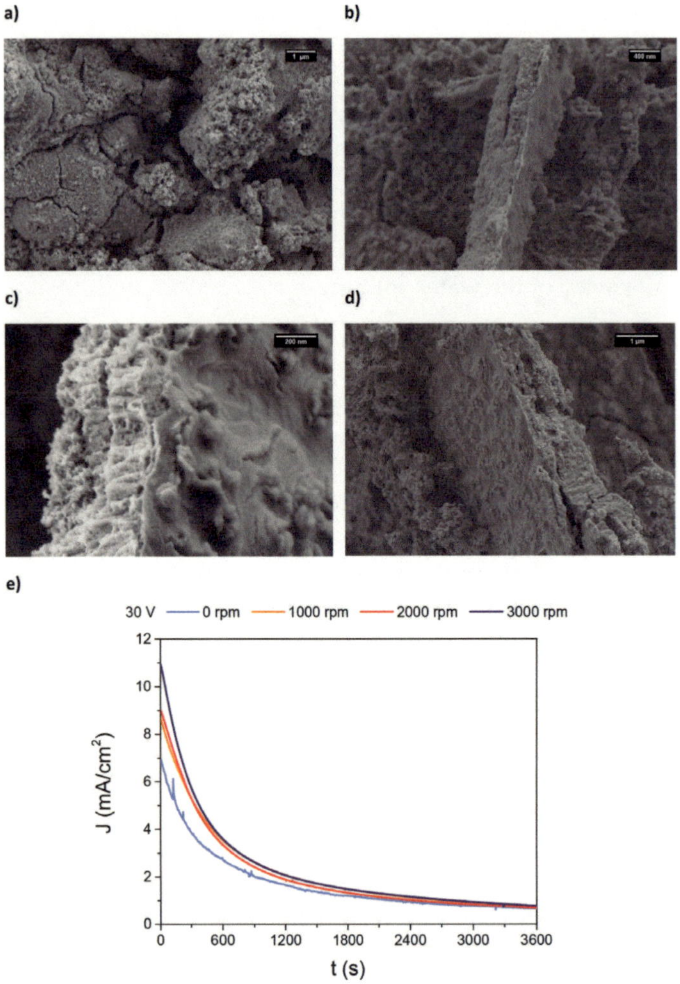

Figure 17. FE-SEM images of the ZnO/ZnS nanostructures anodized at 20 V under static conditions (a), at 1000 rpm (b) and 3000 rpm (c) and at 40 V and under static conditions (d). Current densities recorded during anodization for all the samples anodized at 30 V and at the different rotation speeds (e).

to a porous tubular morphology (**Figure 17c**). Therefore, as rotation speed is increased during anodization, the nanotubular morphology is predominant. This change in morphology continues as anodization potential is increased. In fact, as it can be appreciated in **Figure 17d**, a porous tubular morphology is obtained when anodization is carried out for higher anodization potentials (40 V) even under stagnant conditions. This might be explained taking into consideration that at higher anodization potentials, nanostructures grow faster [75, 76], which involves a lower oxide dissolution, and consequently, a tubular morphology. Besides, the length of the nanostructures was measured obtaining the following values: ~360–600 nm at 20 V, ~800–1100 nm at 30 V and ~1100–1400 nm at 40 V. Thus, longer nanostructures are formed at higher anodization voltages, regardless the hydrodynamic conditions. In this case, longer nanostructures possess a high surface/volume relation, increasing the interaction with light and the electrolyte, and consequently, improving their photocatalytic activity.

5.2. Anodization profile: current densities vs. time

Figure 17e shows, as an example, the current density profiles during the anodization process at 30 V and at the different studied hydrodynamic conditions (0, 1000, 2000 and 3000 rpm). For all cases, the tendency of the current densities is to sharply decrease at the beginning of the anodization, owing to the ZnS precursor layer formation. After that decrease, the current densities continue decreasing until, finally, they stabilize. This tendency is typical of an oxide layer formation in glycerol-based electrolytes [22]. Additionally, it is important to highlight that the higher current density values obtained for the samples anodized stirring the zinc rod are in agreement with the enhancement of the diffusion of the fluoride ions due to the flowing conditions [77]. Besides, the background noise obtained in the current density register of the heterostructure synthesized under static conditions is related to the formation of oxygen bubbles in the zinc surface of the rod during anodization, which generates an irregular anodization of the surface.

It is important to point out that the profiles of current densities versus time for the rest of the heterostructures anodized at 20 and 40 V follow the same trend with hydrodynamic conditions. However, the current densities possess higher values when samples are anodized at higher potentials.

5.3. Crystalline structure characterization

Figure 18a shows, as an example, the Raman spectra of the annealed ZnO/ZnS heterostructures anodized at 30 V at the different hydrodynamic conditions (0, 1000, 2000 and 3000 rpm). Zinc is characterized by a strong broad background signal, which hinders the Raman peaks (**Figure 18a**). This response is associated with the photoluminescence (PL) due to a broad defect band emission in the nanostructures, such as oxygen vacancies [22, 78]. This high concentration of defects was produced due to the absence of oxygen in the atmosphere during the thermal treatment (samples were annealed for 3 h at 375°C in an argon atmosphere), so the evaporated oxygen present in the material cannot be replaced [79] and therefore, the density of oxygen vacancies increases. Then, X-Ray Diffraction (XRD) tests were carried out in order to check the crystalline structure of the annealed samples. **Figure 18b** shows, as an example, the XRD spectrum for the ZnO/ZnS heterostructure anodized at 30 V and at 2000 rpm.

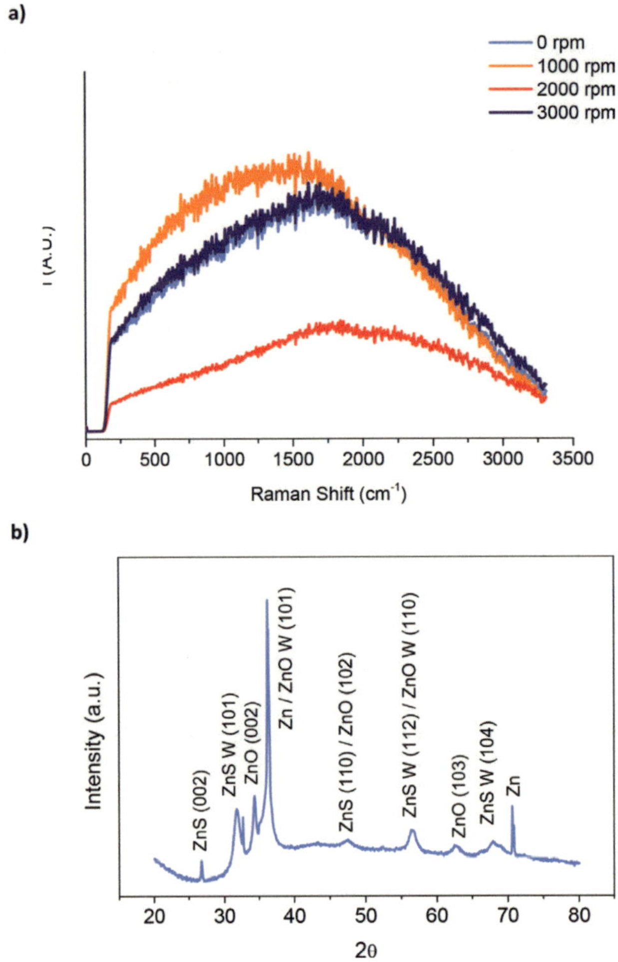

Figure 18. Raman spectra of the ZnO/ZnS heterostructures annealed at 375°C in argon atmosphere for 4 h anodized at 30 V and at 0, 1000, 2000 and 3000 rpm (a) and XRD spectra of the ZnO/ZnS heterostructure anodized at 30 V and 2000 rpm after the annealing treatment.

In the XRD spectra shown in **Figure 18b**, the presence of both ZnO and ZnS can be clearly demonstrated due to the presence of their characteristic peaks [74]. The peak at 25° is related to the ZnS phase with the orientation of (002) [74]. The peaks situated at 36° and 70° are associated with the refraction produced by the polycrystalline Zn substrate [74]. The high intensity of the peak centered at 36° is characteristic of the wurtzite ZnO phase, with an orientation of (101) [80], which is the most abundant phase of the heterostructures. Another ZnO characteristic peak is the one roughly centered at 34°, which corresponds to the orientation (002) [81]. The peaks at 32°, 56° and 67° are associated with the wurtzite ZnS phase with the orientations (101), (112) and (104), respectively [81]. Besides, the peaks centered at 48° and 63° are related, the former, with the ZnS phase with an orientation of (110) and the ZnO phase with an orientation of (102), and the latter, with the ZnO phase with an orientation of (103) [74, 80, 81].

http://dx.doi.org/10.5772/intechopen.78303

5.4. Photoelectrochemical characterization

The ZnO/ZnS nanostructures were characterized as photoanodes in a previously degasified solution of 0.24 M Na_2S and 0.35 M Na_2SO_3 [22]. **Figure 19** shows the photocurrent densities applying different potentials obtained under simulated sunlight AM 1.5 conditions for the different anodization potentials (20, 30 and 40 V) at the studied hydrodynamic conditions (0, 1000, 2000 and 3000 rpm).

In **Figure 19a**, it can be observed that the dark current density values for the ZnO/ZnS heterostructures anodized at 20 V are, in general, high. This fact might be related to a low

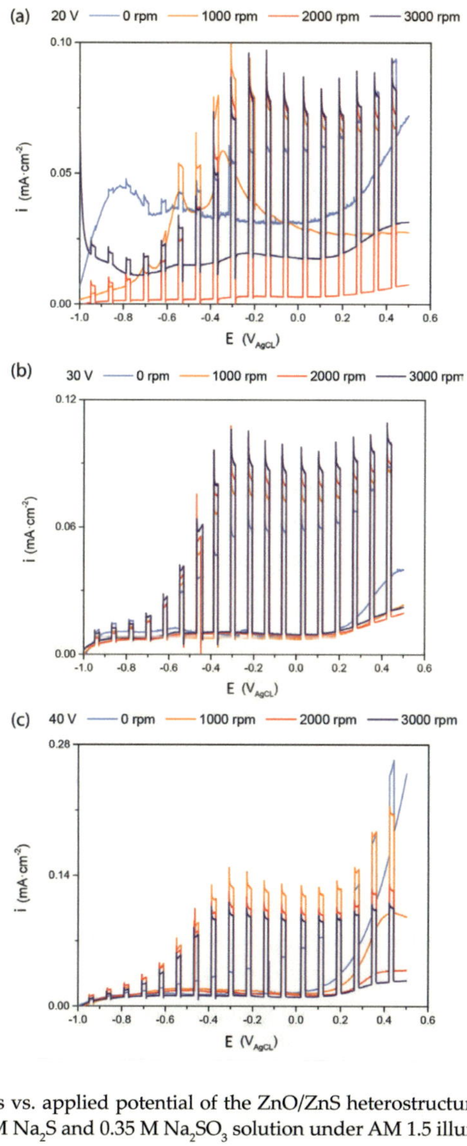

Figure 19. Photocurrent densities vs. applied potential of the ZnO/ZnS heterostructures anodized at 20 V (a), 30 V (b) and 40 V (c) measured in a 0.24 M Na_2S and 0.35 M Na_2SO_3 solution under AM 1.5 illumination.

photocorrosion resistance and could be occasioned by the heterogeneous ZnS precursor layer formed when anodization is performed at 20 V (this was confirmed in FESEM characterization, see **Figure 17a**). On the other hand, when anodization was performed at higher voltages, photocorrosion resistance is increased (**Figure 19b** and **c**), regardless the applied hydrodynamic conditions. In particular, for applied potentials of −0.4 and 0.2 V, the ZnO/ZnS heterostructures show a good stability, since the dark current densities are close to zero and the photocurrent densities are high. Besides, the photocurrent densities for the nanostructures anodized at 30 and 40 V increase for the heterostructures anodized under hydrodynamic conditions. This could be related to the morphology obtained for the nanostructures anodized under hydrodynamic conditions, which showed more nanotubular aspects (**Figure 17**) as the rotation speed was increased, which enhances the charge separation [16, 82–85]. Hence, the photocatalytic response is improved.

Figure 19c shows that the ZnO/ZnS heterostructures anodized at 40 V and under hydrodynamic conditions present the best photocatalytic behavior in terms of higher photocurrent densities and good photocorrosion resistance. In particular, the best conditions were obtained for the ZnO/ZnS heterostructures anodized at 40 V and at 1000 rpm, where the photocurrent density for photoelectrochemical water splitting increased in 100% with respect to the one anodized at 20 V and at 0 rpm, and in a 71% in relation to the photocurrent density obtained for the heterostructure anodized at 20 V and 1000 rpm.

6. Conclusions

In this chapter, the effect of rotating the electrode during electrochemical anodization (hydrodynamic conditions) has been evaluated for different metal oxides. In all cases, hydrodynamic conditions enhanced the photoelectrochemical response of the nanostructures in photoelectrochemical water splitting process.

For nanostructures of TiO_2, hydrodynamic conditions in ethylene glycol media removed the initiation layer over the nanotubes, whereas in glycerol-based electrolytes, rotating the electrode changed the morphology from nanotubes to nanosponges, increasing then the porosity of the structure. For the nanostructures formed in both electrolytes, the surface area increased, resistance values decreased (as EIS showed) and flat band potential became more negative (as MS plots indicated).

In the case of WO_3 nanostructures, the ones synthesized under hydrodynamic conditions presented high active surface area. Furthermore, in the presence of NaF and rotating the electrode, the nanoplatelets grew in a tree-like manner forming globular clusters with more exposed area than in the case of nanostructures synthesized under stagnant conditions. When H_2O_2 is present in the electrolyte, the nanoplatelets were very small, and they aggregated forming very thin layers with high surface areas, which in turn resulted in higher photoelectrochemical efficiencies.

For nanostructures of iron oxide (mainly α-Fe_2O_3), stagnant conditions resulted in nanostructures with an initiation layer that partially covered the top of the nanotubular structure. This layer was removed under hydrodynamic conditions, in particular at 1000 rpm. At higher rotation velocities, collapsed and stacked nanostructures were formed, which caused lower photoelectrochemical efficiencies. Rotating the electrode during anodization at 1000 rpm led to nanotubular structures with suitable donor densities values and flat band potentials (as MS results indicated), which enhanced photoelectrochemical results.

In the case of ZnO/ZnS heterostructures, high applied potentials and hydrodynamic conditions during anodization led to nanotubular morphology which resulted in higher photoelectrochemical performance. In particular, heterostructures anodized at 40 V and under hydrodynamic conditions presented the best photocatalytic behavior and good photocorrosion resistance.

Acknowledgements

The authors would like to express their gratitude for the financial support to the Ministerio of Economía y Competitividad (Projects CTQ2013-42494-R and CTQ2016-79203-R), for its help in the Laser Raman Microscope acquisition (UPOV08-3E-012) and for the co-finance by the European Social Fund.

Author details

Bianca Lucas-Granados, Rita Sánchez-Tovar, Ramón M. Fernández-Domene and José García-Antón*

*Address all correspondence to: jgarciaa@iqn.upv.es

Department Ingeniería Química y Nuclear, Ingeniería Electroquímica y Corrosión (IEC), Universitat Politècnica de Valencia, Valencia, Spain

References

[1] Garcia Bessegato G, Tasso Guaraldo T, Boldrin Zanoni MV. Enhancement of Photoelectrocatalysis efficiency by using nanostructured electrodes. In: Aliofkhazraei M, editor. Modern Electrochemical Methods in Nano, Surface and Corrosion Science. 1st ed. London: IntechOpen; 2014. pp. 271-319

[2] Daghrir R, Drogui P, Robert D. Photoelectrocatalytic technologies for environmental applications. Journal of Photochemistry and Photobiology A. 2012;**238**:41-52

[3] Garcia-Segura S, Brillas E. Applied photoelectrocatalysis on the degradation of organic pollutants in wastewaters. Journal of Photochemistry and Photobiology C. 2017;**31**:1-35

[4] Eftekhari A, Babu VJ, Ramakrishna S. Photoelectrode nanomaterials for photoelectrochemical water splitting. International Journal of Hydrogen Energy. 2017;**42**:11078-11109

[5] Bak T, Nowotny J, Rekas M, Sorrell CC. Photo-electrochemical hydrogen generation from water using solar energy. Materials-related aspects. International Journal of Hydrogen Energy. 2002;**27**:991-1022

[6] Radecka M, Rekas M, Trenczek-Zajac A, Zakrzewska K. Importance of the band gap energy and flat band potential for application of modified TiO_2 photoanodes in water photolysis. Journal of Power Sources. 2008;**181**:46-55

[7] Lide DR, editor. CRC Handbook of Chemistry and Physics. 89th ed. Boca Raton: CRC Press; 2008

[8] Sánchez-Tovar R, Fernández-Domene RM, García-García DM, García-Antón J. Enhancement of photoelectrochemical activity for water splitting by controlling hydrodynamic conditions on titanium anodization. Journal of Power Sources. 2015;**286**:224-231

[9] Roy P, Berger S, Schmuki P. TiO_2 nanotubes: Synthesis and applications. Angewandte Chemie, International Edition. 2011;**50**:2904-2939

[10] van de Krol R, Principles of photoelectrochemical cells: van de Krol R, Grätzel M, editors. Photoelectrochemical Hydrogen Production, 1st ed. New York: Springer; 2012. p. 13-67

[11] Zhang H, Chen G, Bahnemann DW. Photoelectrocatalytic materials for environmental applications. Journal of Materials Chemistry. 2009;**19**:5089-5121

[12] Zheng H, Ou JZ, Strano MS, Kaner RB, Mitchell A, Kalantar-zadeh K. Nanostructured tungsten oxide - properties, synthesis, and applications. Advanced Functional Materials. 2011;**21**:2175-2196

[13] Bignozzi CA, Caramori S, Cristino V, Argazzi R, Meda L, Tacca A. Nanostructured photoelectrodes based on WO_3: Applications to photooxidation of aqueous electrolytes. Chemical Society Reviews. 2013;**42**:2228-2246

[14] Shinde SS, Bhosale CH, Rajpure KY. Photoelectrochemical properties of highly mobilized Li-doped ZnO thin films. Journal of Photochemistry and Photobiology. B. 2013;**120**:1-9

[15] Quintana M, Edvinsson T, Hagfeldt A, Boschloo G. Comparison of dye-sensitized ZnO and TiO2 solar cells - studies of charge transport and carrier lifetime. Journal of Physical Chemistry C. 2007;**111**:1035-1041

[16] Kushwaha A, Aslam M. ZnS shielded ZnO nanowire photoanodes for efficient water splitting. Electrochimica Acta. 2014;**130**:222-231

[17] Zamiri R, Tobaldi DM, Ahangar HA, Rebelo A, Seabra MP, Belsley MS, Ferreira JMF. Study of far infrared optical properties and, photocatalytic activity of ZnO/ZnS hetero-nanocomposite structure. RSC Advances. 2014;**4**:35383-35389

[18] Mor GK, Varghese OK, Paulose M, Grimes CA. Transparent highly ordered TiO_2 nanotube arrays via anodization of titanium thin films. Advanced Functional Materials. 2005;**15**:1291-1296

[19] Tsuchiya H, Macak JM, Sieber I, Taveira L, Ghicov A, Sirotna K, Schmuki P. Self-organized porous WO_3 formed in NaF electrolytes. Electrochemistry Communications. 2005;**7**:295-298

[20] Mor GK, Varghese OK, Paulose M, Shankar K, Grimes CA. A review on highly ordered, vertically oriented TiO_2 nanotube arrays: Fabrication, material properties, and solar energy applications. Solar Energy Materials & Solar Cells. 2006;**90**:2011-2075

[21] Lucas-Granados B, Sánchez-Tovar R, Fernández-Domene RM, García-Antón J. Study of the annealing conditions and photoelectrochemical characterization of a new iron oxide bi-layered nanostructure for water splitting. Solar Energy Materials & Solar Cells. 2016;**153**:68-77

[22] Sanchez-Tovar R, Fernandez-Domene RM, Montañés MT, Sanz-Marco A, Garcia-Anton J. ZnO/ZnS heterostructures for hydrogen production by photoelectrochemical water splitting. RSC Advances. 2016;**6**:30425-30435

[23] Fernández-Domene RM, Sánchez-Tovar R, Lucas-Granados B, García-Antón J. Improvement in photocatalytic activity of stable WO_3 nanoplatelet globular clusters arranged in a tree-like fashion: Influence of rotation velocity during anodization. Applied Catalysis B: Environmental. 2016;**189**:266-282

[24] Shankar K, Mor GK, Prakasam HE, Yoriya S, Paulose M, Varghese O, Grimes CA. Highly-ordered TiO_2 nanotube arrays up to 220 mm in length: Use in water photoelectrolysis and dye-sensitized solar cells. Nanotechnology. 2007;**18**:065707 (11pp)

[25] Bak T, Nowotny J, Rekas M, Sorrell CC. Photo-electrochemical properties of the TiO_2-Pt system in aqueous solutions. International Journal of Hydrogen Energy. 2002;**27**:19-26

[26] Yahia SAA, Hamadou L, Kadri A, Benbrahim N, Sutter EMM. Effect of anodizing potential on the formation and EIS characteristics of TiO_2 nanotube arrays. Journal of the Electrochemical Society. 2012;**159**:K83-K92

[27] Sánchez-Tovar R, Lee K, García-Antón J, Schmuki P. Formation of anodic TiO_2 nanotube or nanosponge morphology determined by the electrolyte hydrodynamic conditions. Electrochemistry Communications. 2013;**26**:1-4

[28] Costa LL, Prado AGS. TiO_2 nanotubes as recyclable catalyst for efficient photocatalytic degradation of indigo carmine dye. Journal of Photochemistry and Photobiology A: Chemistry. 2009;**201**:45-49

[29] Hsiao PT, Wang KP, Cheng CW, Teng HS. Nanocrystalline anatase TiO_2 derived from a titanate-directed route for dye-sensitized solar cells. Journal of Photochemistry and Photobiology A. 2007;**188**:19-24

[30] Qian L, Du ZL, Yang SY, Jin ZS. Raman-study of titania nanotube by soft chemical-process. Journal of Molecular Structure. 2005;**749**:103-107

[31] Hirschorn B, Orazem ME, Tribollet B, Vivier V, Frateur I, Musiani M. Determination of effective capacitance and film thickness from constant-phase-element parameters. Electrochimica Acta. 2010;**55**:6218-6227

[32] Palmas S, Polcaro AM, Ruiz JR, Da Pozzo A, Mascia M, Vacca A. TiO_2 photoanodes for electrically enhanced water splitting. International Journal of Hydrogen Energy. 2010;**35**:6561-6570

[33] Aïnouche L, Hamadou L, Kadri A, Benbrahim N, Bradai D. Interfacial barrier layer properties of three generations of TiO_2 nanotube arrays. Electrochimica Acta. 2014;**133**:597-609

[34] Jiang Z, Dai X, Middleton H. Investigation on passivity of titanium under steady-state conditions in acidic solutions. Materials Chemistry and Physics. 2011;**126**:859-865

[35] Kong DS, Lu WH, Feng YY, Yu ZY, Wu JX, Fan WJ, Liu HY. Studying on the point-defect-conductive property of the semiconducting anodic oxide films on titanium. Journal of the Electrochemical Society. 2009;**156**:C39-C44

[36] Sazou D, Saltidou K, Pagitsas M. Understanding the effect of bromides on the stability of titanium oxide films based on a point defect model. Electrochimica Acta. 2012;**76**:48-61

[37] Roh B, Macdonald DD. Effect of oxygen vacancies in anodic titanium oxide films on the kinetics of the oxygen electrode reaction. Russian Journal of Electrochemistry. 2007; **43**:125-135

[38] Peng H. First-principles study of native defects in rutile TiO_2. Physics Letters A. 2008;**372**:1527-1530

[39] Tsui LK, Homma T, Zangari G. Photocurrent conversion in anodized TiO_2 nanotube arrays: Effect of the water content in anodizing solutions. Journal of Physical Chemistry C. 2013;**117**:6979-6989

[40] Radecka M, Wierzbicka M, Komornicki S, Rekas M. Influence of Cr on photoelectrochemical properties of TiO_2 thin films. Physica B: Condensed Matter. 2004;**348**:160-168

[41] Carp O, Huisman CL, Reller A. Photoinduced reactivity of titanium dioxide. Progress in Solid State Chemistry. 2004;**32**:33-177

[42] Wang D, Zhang X, Sun P, Lu S, Wang L, Wang C, Liu Y. Photoelectrochemical water splitting with rutile TiO_2 nanowires array: Synergistic effect of hydrogen treatment and surface modification with anatase nanoparticles. Electrochimica Acta. 2014;**130**:290-295

[43] Morgan BJ, Watson GW. Polaronic trapping of electrons and holes by native defects in anatase TiO_2. Physical Review B. 2009;**80**:233102

[44] Irie H, Watanabe Y, Hashimoto K. Nitrogen-concentration dependence on photocatalytic activity of $TiO_{2-x}N_x$ powders. The Journal of Physical Chemistry. B. 2003;**107**:5483-5486

[45] de Tacconi NR, Chenthamarakshan CR, Yogeeswaran G, Watcharenwong A, de Zoysa RS, Basit NA, Rajeshwar K. Nanoporous TiO_2 and WO_3 films by anodization of titanium and tungsten substrates: Influence of process variables on morphology and photoelectrochemical response. The Journal of Physical Chemistry. B. 2006;**110**:25347-25355

[46] Nah YC, Ghicov A, Kim D, Schmuki P. Enhanced electrochromic properties of self-organized nanoporous WO_3. Electrochemistry Communications. 2008;**10**:1777-1780

[47] Lai CW, Sreekantan S. Fabrication of WO_3 nanostructures by anodization method for visible-light driven water splitting and photodegradation of methyl orange. Materials Science in Semiconductor Processing. 2013;**16**:303-310

[48] Ou JZ, Balendhran S, Field MR, McCulloch DG, Zoolfakar AS, Rani RA, Zhuiykov S, O'Mullane AP, Kalantar-zadeh K. The anodized crystalline WO_3 nanoporous network with enhanced electrochromic properties. Nanoscale. 2012;**4**:5980-5988

[49] Ou JZ, Rani RA, Balendhran S, Zoolfakar AS, Field MR, Zhuiykov S, O'Mullane AP, Kalantar-zadeh K. Anodic formation of a thick three-dimensional nanoporous WO_3 film and its photocatalytic property. Electrochemistry Communications. 2013;**27**:128-132

[50] Reyes-Gil KR, Wiggenhorn C, Brunschwig BS, Lewis NS. Comparison between the quantum yields of compact and porous WO_3 photoanodes. Journal of Physical Chemistry C. 2013;**117**:14947-14957

[51] Liu Y, Li Y, Li W, Han S, Liu C. Photoelectrochemical properties and photocatalytic activity of nitrogen-doped nanoporous WO_3 photoelectrodes under visible light. Applied Surface Science. 2012;**258**:5038-5045

[52] Nah YC, Paramasivam I, Hahn R, Shrestha NK, Schmuki P. Nitrogen doping of nanoporous WO_3 layers by NH_3 treatment for increased visible light photoresponse. Nanotechnology. 2010;**21**:105704 (5pp)

[53] Sadek AZ, Zheng H, Breedon M, Bansal V, Bhargava SK, Latham K, Zhu J, Yu L, Hu Z, Spizzirri PG, Wlodarski W, Kalantar-zadeh K. High-temperature anodized WO_3 nanoplatelet films for photosensitive devices. Langmuir. 2009;**25**:9545-9551

[54] Kalantar-zadeh K, Sadek AZ, Zheng H, Bansal V, Bhargava SK, Wlodarski W, Zhu J, Yu L, Hu Z. Nanostructured WO_3 films using high temperature anodization. Sensors and Actuators B: Chemical. 2009;**142**:230-235

[55] Ng C, Ng YH, Iwase A, Amal R. Influence of annealing temperature of WO_3 in Photoelectrochemical conversion and energy storage for water splitting. ACS Applied Materials & Interfaces. 2013;**5**:5269-5275

[56] Cristino V, Caramori S, Argazzi R, Meda L, Marra GL, Bignozzi CA. Efficient photoelectrochemical water splitting by anodically grown WO_3 electrodes. Langmuir. 2011;**27**:7276-7284

[57] Ng C, Ye C, Ng YH, Amal R. Flower-shaped tungsten oxide with inorganic fullerene-like structure: Synthesis and characterization. Crystal Growth & Design. 2010;**10**:3794-3801

[58] Fernández-Domene RM, Sánchez-Tovar R, Segura-Sanchís E, García-Antón J. Novel tree-like WO_3 nanoplatelets with very high surface area synthesized by anodization under controlled hydrodynamic conditions. Chemical Engineering Journal. 2015;**286**:59-67

[59] Fernández-Domene RM, Sánchez-Tovar R, Lucas-Granados B, Roselló-Márquez G, García-Antón J. A simple method to fabricate high-performance nanostructured WO_3 photocatalysts with adjusted morphology in the presence of complexing agents. Materials and Design. 2017;**116**:160-170

[60] Li W, Li J, Wang X, Luo S, Xiao J, Chen Q. Visible light photoelectrochemical responsiveness of self-organized nanoporous WO_3 films. Electrochimica Acta. 2010;**56**:620-625

[61] Wang CK, Lin CK, Wu CL, Wang SC, Huang JL. Synthesis and characterization of electrochromic plate-like tungsten oxide films by acidic treatment of electrochemical anodized tungsten. Electrochimica Acta. 2013;**112**:24-31

[62] Lassner E, Schubert WD. Tungsten. Properties, Chemistry, Technology of the Element, Alloys, and Chemical Compounds. 1st ed. New York: Kluwer Academic/Plenum Publishers; 1999

[63] Lucas-Granados B, Sánchez-Tovar R, Fernández-Domene RM, García-Antón J. Influence of electrolyte temperature on the synthesis of iron oxide nanostructures by electrochemical anodization for water splitting. International Journal of Hydrogen Energy. 2018;**43**:7923-7937

[64] Pervez SA, Kim D, Farooq U, Yaqub A, Choi JH, Lee YJ. Crystalline iron oxide nanotube arrays with high aspect ratio as binder free anode for Li-ion batteries. Physica Status Solidi a - Applications and Materials Science. 2014;**211**:1889-1894

[65] Xie K, Guo M, Huang H, Liu Y. Fabrication of iron oxide nanotube arrays by electrochemical anodization. Corrosion Science. 2014;**88**:66-75

[66] Yasuda K, Schmuki P. Control of morphology and composition of self-organized zirconium titanate nanotubes formed in $(NH_4)_2SO_4/NH_4F$ electrolytes. Electrochimica Acta. 2007;**52**:4053-4061

[67] Macak JM, Hildebrand H, Marten-Jahns U, Schmuki P. Mechanistic aspects and growth of large diameter self-organized TiO2 nanotubes. Journal of Electroanalytical Chemistry. 2008;**621**:254-266

[68] Lee CY, Wang L, Kado Y, Killian MS, Schmuki P. Anodic nanotubular/porous hematite photoanode for solar water splitting: Substantial effect of iron substrate purity. ChemSusChem. 2014;**7**:934-940

[69] Lucas-Granados B, Sánchez-Tovar R, Fernández-Domene RM, García-Antón J. Controlled hydrodynamic conditions on the formation of iron oxide nanostructures synthesized by electrochemical anodization: Effect of the electrode rotation speed. Applied Surface Science. 2017;**392**:503-513

[70] a de Faria DL, Lopes FN. Heated goethite and natural hematite: Can Raman spectroscopy be used to differentiate them. Vibrational Spectroscopy. 2007;**45**:117-121

[71] Jubb AM, Allen HC. Vibrational spectroscopic characterization of hematite, maghemite, and magnetite thin films produced by vapor deposition. ACS Applied Materials & Interfaces. 2010;**2**:2804-2812

[72] Rahman G, Joo OS. Photoelectrochemical water splitting at nanostructured α-Fe_2O_3 electrodes. International Journal of Hydrogen Energy. 2012;**37**:13989-13997

[73] Shen S, Zhou J, Dong CL, Hu Y, Tseng EN, Guo P. Surface engineered doping of hematite nanorod arrays for improved photoelectrochemical water splitting. Scientific Reports. 2014;**4**:6627-6636

[74] Shrestha NK, Lee K, Hahn R, Schmuki P. Anodic growth of hierarchically structured nanotubular ZnO architectures on zinc surfaces using a sulfide based electrolyte. Electrochemistry Communications. 2013;**34**:9-13

[75] Zhou X, Truong Nguyen N, Özkan S, Schmuki S. Anodic TiO_2 nanotube layers:Why does self-organized growth occur — A mini review. Electrochemistry Communications. 2014;**46**:157-162

[76] Khudhair D, Bhatti A, Li Y, Amani Hamedani H, Garmestani H, Hodgson P, Nahavandi S. Anodization parameters influencing the morphology and electrical properties of TiO_2 nanotubes for living cell interfacing and investigations. Materials Science and Engineering: C. 2016;**59**:1125-1142

[77] Sánchez-Tovar R, Fernández-Domene RM, Martínez-Sánchez A, Blasco-Tamarit E, García-Antón J. Synergistic effect between hydrodynamic conditions during Ti anodization and acidic treatment on the photoelectric properties of TiO_2 nanotubes. Journal of Catalysis. 2015;**330**:434-441

[78] He S, Zheng M, Yao L, Yuan X, Li M, Mab L, Shen W. Preparation and properties of ZnO nanostructures by electrochemical anodization method. Applied Surface Science. 2010;**256**:2557-2562

[79] Ogata K, Sakurai K, Fujita S, Fujita S, Matsushige K. Effects of thermal annealing of ZnO layers grown by MBE. Journal of Crystal Growth. 2000;**214/215**:312-315

[80] Farooqi MMH, Srivastava RK. Structural, optical and photoconductivity study of ZnO nanoparticles synthesized by annealing of ZnS nanoparticles. Journal of Alloys and Compounds. 2017;**691**:275-286

[81] Patel PC, Ghosh S, Srivastava PC. Structural, magnetic and optical properties of ZnO nanostructures converted from ZnS nanoparticles. Materials Research Bulletin. 2016;**81**:85-89

[82] Chen S, Thind SS, Chen A. Nanostructured materials for water splitting - state of the art and future needs: A mini-review. Electrochemistry Communications. 2016;**63**:10-17

[83] Zhu J, Zäch M. Nanostructured materials for photocatalytic hydrogen production. Current Opinion in Colloid & Interface Science. 2009;**14**:260-269

[84] Kim Y, Park J, Kim S, Park DW, Choi J. Fabrication of hierarchical ZnO nanostructures for dye-sensitized solar cells. Electrochimica Acta. 2012;**78**:417-421

[85] Sreekantan S, Gee LR, Lockman Z. Room temperature anodic deposition and shape control of one-dimensional nanostructured zinc oxide. Journal of Alloys and Compounds. 2009;**476**:513-518

Chapter 5

Nanostructured Perovskites for Catalytic Combustion

Corneliu Doroftei

Additional information is available at the end of the chapter

http://dx.doi.org/10.5772/intechopen.77727

Abstract

Until recently, for reducing the environment pollution, the combustion catalysts based on noble metals was considered the most active, practically irreplaceable. Their high cost, high thermal instability, high sensitivity to deactivation and to the attack of some harmful elements or compounds determine the intensification of studies to replace them with new cheaper and stable catalysts. Numerous experimental data from literature indicate that the semiconductive oxidic compounds can compete with combustion catalysts based on noble metals from catalytic activity standpoint. Recent studies led to the realization of remarkable catalytic activity at moderate (350–600°C) and high (over 600°C) temperatures at some oxidic perovskite compounds, which contain transition metals. In this study are presented a series of nanostructured oxidic compounds with perovskite structure, based on transition metals and synthesized by the precursor method of self-combustion, using polyvinyl alcohol as colloidal medium, for catalytic combustion of some volatile organic compounds at low (50–350°C) and moderate temperatures. The catalytic activity of the perovskite compounds in the total oxidation reactions of the gases is largely determined by the amount of weakly bound surface oxygen species which in turn depends on the presence of oxygen vacancies.

Keywords: perovskites, nanostructures, oxidation, self-combustion, catalytic combustion

1. Introduction

Until recently, for reducing the environment pollution, the combustion catalysts based on noble metals were considered the most active, practically irreplaceable. Their high cost, high thermal instability, high sensitivity to deactivation and to the attack of some harmful elements or compounds determine the intensification of studies to replace them with new cheaper and stabler catalysts.

Catalysis is a procedure through which the kinetic of a chemical reaction is positively or negatively influenced by the presence of another substance, called catalyst. The catalyst is a substance, which participates in reaction, but is found unchanged at the end of reaction. The catalyst does not appear in the equation of the chemical reaction, nor is it in an apparent stoichiometric ratio with the reacting substances.

Catalyst important property consists of that it can reduce the reaction activation energy. In this way, the reactions can be triggered and developed in such conditions of temperature, pressure or concentration at which the reaction could not occur without catalysis. The catalyst can only accelerate or trigger the thermodynamically possible reactions, namely spontaneous reactions running freely toward the equilibrium, therefore reactions with negative free enthalpy.

Catalysts can be metals, oxides, bases, salts, and so on, which once introduced in the reaction medium in small or very small amounts, take part in a certain stage of chemical reaction, but at the end of the reaction, they can be found in the initial quantity without suffering any chemical transformation.

Catalysis can be of two types:

- Homogeneous catalysis in which both the catalyst and the reacting substances are miscible and form a homogeneous system consisting of a single phase, gaseous or liquid, for instance an acid added in an aqueous solution of reagents. Catalyst intervention is exclusively chemical.
- Heterogeneous catalysis, in which the catalysts and the reagents form a heterogeneous system consisting of several phases. The catalyst is usually under solid state, while the reagents are in solid or gaseous state. Catalyst intervention is both chemical and physical.

Heterogeneous catalysis occurs in three successive stages:

- adsorption of reagents molecules on catalyst surface;
- reaction between adsorbed molecules, with the participation of catalyst active centers;
- desorption of reaction products, that is, the substances resulted from reaction pass into the environmental medium.

Adsorption can be of two kinds: adsorption though van der Waals forces, and activated adsorption (chemosorption). Adsorption though van der Waals forces occurs rapidly and at low temperatures, and the van der Waals heats are small. Activated adsorption (chemosorption) is slow and accompanied by a bigger heat elimination than in the previous case because bonds of chemical nature are also established, as the result of the interaction between reagents molecules and the atoms of the catalyst active centers.

Combustion (burning) of a gas in air is a homogeneous reaction at high temperature and occurs in certain conditions [1]. The air/gas ratio has a minimum value and a maximum value (defining the flammability interval) within which the combustion can occur, interval specific to each flammable gas. Flammability interval increases with the increase of temperature and decrease of activation energy.

At a high enough temperature or at a small enough activation energy, combustion can occur for any air/gas ratio. Burning in the air the most usual combustibles, hydrocarbons, is accompanied by the release of polluting gases, especially carbon monoxide (CO), nitrogen oxides (NO_x) and various unburnt hydrocarbons (HC). **Figure 1** shows the variation of polluting gas concentration (pollution level) with temperature during hydrocarbons burning in air. At higher burning temperatures, the concentration of nitrogen oxides increases. With the decrease of burning temperature, the amount of carbon monoxide and unburnt hydrocarbon increases. One can state that burning any combustible in air is unavoidably accompanied by polluting gases generation.

Catalytic combustion is the reaction of combustible gas oxidation in the presence of a catalyst. In almost all the cases, it is an heterogeneous reaction, the catalyst being a solid substance. Depending on catalyst, the reaction can occur at low, medium or high temperature, at very small concentrations of gas in air, and can be complete without generating polluting gases.

Unlike the noncatalytic combustion, at which the reaction thermally propagates throughout the available volume, in the case of catalytic combustion, the reaction only occurs at the catalyst surface. The model of the catalytic combustion describes a flux of reagents molecules toward the catalyst surface, and a reverse flux of reaction products molecules, as well as a heat flux. Catalytic combustion can be amplified by increasing the catalyst surface or the gas velocity at catalyst surface.

Usually, the catalyst consists of a substrate (metallic or ceramic) which provides the catalyst its form and mechanical strength, a ceramic support (alumina, zirconia, magnesia, silica, ceria, hexa-aluminates, etc.) with high specific surface area (aria/mass), and an active substance impregnated in support or deposited on it [2–4]. Sometimes, the active atoms are also introduced in the oxide composition support [5].

Catalytic combustion can occur at low (between 50 and 350°C), medium (between 350 and 600°C), or high (above 600°C) temperature.

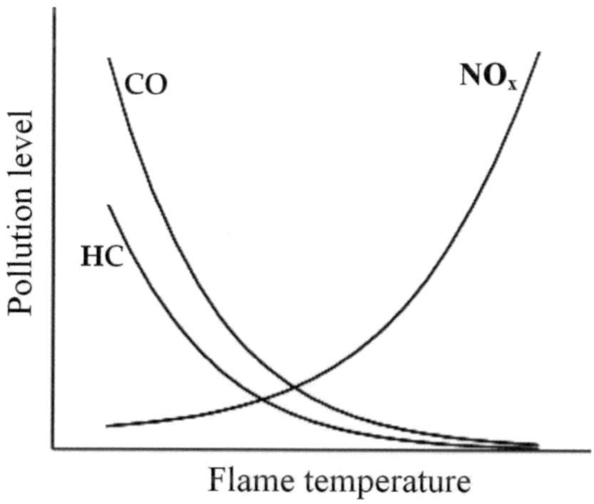

Figure 1. Variation of the pollution level with temperature during hydrocarbons burning in air.

Catalytic combustion at low and medium temperature is used to eliminate from air the polluting gases, volatile organic compounds (VOC), such as solvents vapors, mine gases, combustible gases released from industrial installations, and so on. Combustion occurs according to usual reactions, such as:

$$2 \cdot CO + O_2 \rightarrow 2 \cdot CO_2 \tag{1}$$

$$(C_x H_y O_z) + O_2 \rightarrow CO_2 + H_2O \tag{2}$$

Generally, it is used when the concentration of combustible gases in air is very reduced and unable to keep the fire alive [6]. Still another possible utilization now studied is at combustible cells that can work at such temperatures. In this case, catalysts efficient at low and medium temperature which significantly reduces the activation energy of combustion reactions are used.

High-temperature catalytic combustion pursues complete burning and diminution of polluting gas emission. It is generally used in the installations which generate heat and energy. The concentration of combustible gases is high enough for a homogeneous combustion, but the purpose is a complete burning and diminution of polluting gases. The diminution of activation energy results in catalytic combustion of combustible polluting gases, while the decrease of the burning temperature decreases the concentration of nitrogen oxides. Through this type of catalysis, one can produce stable burning of combustible gases with small caloric power, different types of biogases, as well as of some gases considered as noncombustible, such as ammonia, chlorinated hydrocarbons or nitrogen monoxide, like in the reactions below:

$$4 \cdot NH_3 + 3 \cdot O_2 \rightarrow 2 \cdot N_2 + 6 \cdot H_2O \tag{3}$$

$$(C_x H_y Cl_z) + O_2 \rightarrow CO_2 + H_2O + HCl \tag{4}$$

$$4 \cdot NO + 4 \cdot NH_3 + O_2 \rightarrow 4 \cdot N_2 + 6 \cdot H_2O \tag{5}$$

The utilized catalysts in this case are from the category of those stable at high temperature, which yet are not the most active. For this reason, usually a number of catalysts are used in several stages of the combustion device temperature.

A special category of catalytic combustion is that used to reduce the pollutants from burnt gases evacuated by gas turbine and car motors (catalytic post combustion). The reactions occur in difficult conditions: moderate temperature, reduced oxygen amount, big number of polluting compounds in small concentration. A combination of catalysts is used, having both a reducing character to reduce the nitrogen oxides, and oxidizing character, for unburnt hydrocarbons oxidation [7]. There is also a mechanism of nitrogen oxides catalytic reduction by the combustible gases (unburnt hydrocarbons and carbon monoxide).

Until the recent years, aiming to reduce the environment pollutants, catalysts based on noble metals (platinum, gold, palladium, rhodium, iridium), as well as their alloys with other metals have been considered the most active, practically irreplaceable. Yet, their high cost, high temperature instability, high sensitivity to deactivation and to noxious elements or compounds, determined an intensification of the studies meant to replace them with cheaper and more stable catalysts [2, 8, 9]. Due to the high cost of the noble metals, the utilization of these catalysts is restricted to industrial processes where the noble metals from used catalysts are recovered. The literature indicates that the oxide compounds can enter in competition with catalysts based on noble metals in terms of catalytic activity. Remarkable catalytic properties have been discovered to a number of oxide compounds: oxides of some metals, ferrites, perovskites, and so on. Recent researches resulted in the realization of remarkable catalytic activities at medium and low temperatures at some oxide compounds with perovskite structure. The first practical results were reported more than two decades ago, but most of the researches were performed in the last years.

Simple perovskites have the general formula ABO_3, the same as the natural mineral "perovskite" ($CaTiO_3$) with face centered cubic crystalline structure, but the perovskites can also have other crystalline structure [10]. The elementary cell of a simple perovskite with cubic structure consists of a single molecule. In the peaks of the cube there are the A type ions with big ionic radius, which can be, for instance, Ca^{2+}, Ba^{2+}, Sr^{2+}, Cd^{2+}, Pb^{2+}, La^{3+}, Pr^{3+}, Nd^{3+}, Gd^{3+}, Y^{3+}. In the cube center, there are B-type ions with smaller radius, for example Fe^{3+}, Mn^{3+}, Cr^{3+}, Al^{3+}, Ti^{4+}, Mn^{4+}. The oxygen ions are located in the cube face centers. The sample surface properties (surface composition and specific area) are also important in the case of catalytic combustion that uses perovskites.

Porous nanocrystalline compounds have a quite large surface area. Nanocrystalline particles can have a normal crystalline core and a shell with altered crystalline structure, their composition being different from that of the core. Their properties can be controlled through substitutions at A and B positions or by modifying the oxygen quantity. Through a slight modification of the initial composition of perovskite material, following a heat treatment, the dopant ions are concentrated at particles surface and influence to a great extent the physical surface properties.

A catalyst good for catalytic combustion of gases or vapors must satisfy four major conditions:

- reduce the combustible gas concentration to minimum;
- activate at minimum values of gas concentration;
- manifest catalytic conversion at minimum temperature;
- have a maximum utilization duration until deactivation (poisoning).

Since the heterogeneous catalysis is a surface phenomenon, its efficiency is determined by both chemical composition and catalyst surface structure. Generally, nanostructured oxide compounds are used [2, 11–13] obtained by various procedures [11, 12, 14, 15]. The nanostructure is the one which provides a large specific surface area and a superior catalyst reactivity [15]. Generally, a nanoparticle possesses a single phase crystalline core and a shell with altered crystalline structure and with several secondary phases, having a composition different from that of the core. When the material composition is non-stoichiometric, the atoms in excess are concentrated into the phases which form the particle shell that takes effective part in catalysis. Particles shape, dimension and superficial structure are determined by their preparation procedure. If the oxide nanoparticles are obtained through special synthesis and growth procedures, they have special catalytic properties which depend on the preparation procedure and the used support [11, 16].

There are numerous procedures to produce nanoparticles of oxide compounds, some general and other specific to catalysts. A procedure specific for catalysts preparation is support impregnation. The support is realized under the form of a ceramic microporous material and is impregnated with salts solutions [4, 7]. By assembly calcination, oxide compounds with nanostructure are formed inside the pores. The catalytic reaction occurs inside the pores and is intensified by a forced gas circulation through the catalyst. Preparation procedures for nanoparticles with catalytic properties represent the object of numerous researches and invention patents. The composition of combustion oxide catalysts is very diversified. The substances are stable oxide compounds and contain various active metal ions. Utilization of nontoxic compounds, easy to prepare, with high thermal and chemical stability and a low cost is preferred.

2. Reports from the specialized literature on a series of perovskites used for catalysts realization

Systematic studies of a big number of oxide compounds have proved that the catalytic activity represents a common phenomenon for the oxide compounds, perovskites included [17–30]. Recent specialized literature reports on a series of perovskites used for catalysts realization.

One of the first studied oxide catalysts [17] was simple perovskite $CuCoO_3$ on γ-Al_2O_3 support. This has been analyzed in comparison with a classical catalyst with platinum, in propylene elimination from the gases released in polypropylene preparation installations. It was found that this catalyst can successfully replace the classical catalyst with platinum.

Milt et al. have prepared perovskites $LaCoO_3$ and $LaFeO_3$ by the explosion method obtaining methane catalytic conversion over 650°C [18].

Hammami et al. obtain simple lanthanum $LaMnO_{3+y}$ prepared by citric acid gel process and heating at 700°C obtaining methane catalytic conversion over 600°C [19].

The complex perovskite with oxygen deficit $La_{1-x}A_xFeO_{3-\delta}$, (where A = Ca, Sr. and x = 0.1–0.2) prepared through the method of complexing with citric acid was studied for catalytic combustion of ammonia at temperatures of 700–1000°C. A good catalyst selectivity to the reaction of nitrogen oxide formation was found [20].

Batis et al. obtain good catalytic properties in methane combustion over 800°C using manganite $La_{0.4}Ca_{0.6}MnO_{3.03}$ prepared by thermal decomposition of a liquid precursor of lanthanum and calcium nitrates and manganese acetate, followed by calcination [21]. Ponce et al. found that the stability of Mn^{4+} ions seems to be the crucial factor determining the catalytic activity of manganites in the oxidation of methane in the temperature range 200–800°C [22].

Alifanti et al. synthesized $LaCoO_3$ perovskite on $Ce_{1-x}Zr_xO_2$ (x = 0–0.3), for which the support was prepared by complexation with citric acid and calcined at 700°C for 6 h. The support was successively impregnated with excess aqueous solution of La and Co nitrates (La/Co ratio 1/1) and citric acid, and the surface area achieved was between 23.6 and 22.7 $m^2 g^{-1}$ for the 10 and 20 wt% supported catalysts, respectively, and 11.3 $m^2 g^{-1}$ for the unsupported perovskite. The same study reported that $LaCoO_3/Ce_{0.9}Zr_{0.1}O_2$ (20 wt%) achieved the best performances completely oxidizing 1700 ppm of benzene and toluene (air flow 100 mL/min) at around 450 and 350°C, respectively. It also found that the bulk perovskite achieved a total conversion of benzene at a similar temperature while a higher temperature of 450°C was needed to fully oxidize toluene [23, 24].

Zhang et al. realized a study on some perovskites type La–Sr–Co–O, being able to obtain in the case of the compound $La_{0.7}Sr_{0.3}CoO_3$ the catalytic combustion of CO over 125°C [25].

Huang et al. prepared $La_{0.8}Cu_{0.2}MnO_3$ and $La_{0.8}Sr_{0.2}MnO_3$ for the removal of toluene (6000 ppm GHSV 5000 h^{-1}) in the presence of dodecyl mercaptan and both catalysts lost activity over time due to the formation of $CuSO_4$ or $SrSO_4$. The activity decreased with the increase of SO_2 concentration and dropped to around 30 and 20% of toluene conversion over 150 h operations with SO_2 of 60 ppm for $La_{0.8}Sr_{0.2}MnO_3$ and $La_{0.8}Cu_{0.2}MnO_3$, respectively [23, 26].

3. Obtaining nanomaterials with perovskite structure by the precursor method of self-combustion

A recent method of nanoparticles preparation, including those with perovskite structure is the precursor method of self-combustion [31–38], which uses as colloidal precipitation medium the polyvinyl alcohol in various mass ratios with the metals. This method permits to obtain oxide compounds with homogeneous nanograined porous highly pure structure and reproducible properties. It is superior to other known conventional methods.

In essence, the precursor method of self-combustion consists of the co-precipitation with ammonia of the hydroxides from a mixture of nitrates solutions in a colloidal medium of polyvinyl alcohol. The dried precipitate contains an intimate and very homogeneous mixture of hydroxides, ammonium nitrate and polyvinyl alcohol. Microstructure of the precipitate and the final product can be modified by changing the ratio between the mass of polyvinyl alcohol and the mass of metal ions [32]. As in the case of a pyrotechnical mixture, the dry precipitate is enflamed at a point and suffers a rapid autonomous combustion which results, due to the very short reaction time, an oxide compound with nanometer structure, under the shape of powder (**Figure 2(a)**). A subsequent heat treatment will result in a controlled increase

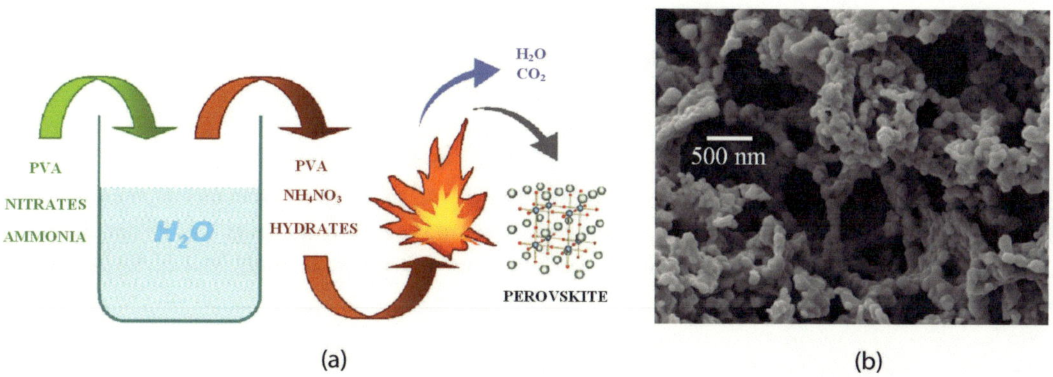

Figure 2. Getting perovskites by the precursor method of self-combustion (a); SEM micrography for a perovskite obtained by this method (b).

of crystallites to the size necessary for applications, with single phase core and a possible poly-phase superficial layer.

In the SEM micrography of a sample of LPMO perovskite (**Figure 2(b)**), one can notice agglomerations of sub-micron grains of irregular shape, with inter-grains pores and large tubular pores (under 1 μm), favorable to gas access into the samples. This open-pores system appeared during the self-combustion reaction, through which a large amount of gases was eliminated. This porous structure appears at all the oxidic compounds prepared through this method.

4. Nanostructured perovskites for catalytic combustion of gases and of some volatile organic compounds

In this subsection are presented and characterized a series of perovskites ($SrMnO_3$, $SrCoO_3$, $MnFeO_3$ and $GdAlO_3$) obtained by the precursor method of self-combustion use for catalytic combustion of some volatile organic compounds (acetone, benzene, propane and Pb free gasoline).

4.1. Catalysts preparation and characterization

Nanograined perovskite powders of nominal compositions: $SrMnO_3$ (P1), $SrCoO_3$ (P2), $MnFeO_3$ (P3) and $GdAlO_3$ (P4) were prepared by the precursor method of self-combustion followed by heat treatment. It was used metal nitrates, ammonium hydroxide and polyvinyl alcohol (PVA) as starting materials. A solution containing metal nitrates was mixed with an aqueous solution (10% concentration) of polyvinyl alcohol. A small amount of NH_4OH solution (10% concentration) was dropped to adjust the pH value to about 8. This produced a sol of metal hydroxides and ammonium nitrate. By drying at 100°C for 12 h, the sol was turned into a dried gel. The dried gel was ignited in a corner and a combustion reaction spontaneously propagated through the whole gel. The obtained powders were calcinated at 500°C to eliminate the residual organic compounds. Finally, the calcined powders were annealed in air at 1000°C, 7 h.

The structure and surface properties of the heat-treated samples were investigated by X-ray diffraction (XRD), scanning electron microscopy (SEM), energy dispersive X-ray analysis (EDX) and Brunauer-Emmett-Teller (BET) analysis.

The catalytic testing of the perovskite catalysts in the flameless combustion of some selected gases was carried out at atmospheric pressure in a flow-type set-up (flow rate of 100 cm^3/min and VOC concentration in air of 1–2‰). The catalyst powder (0.3–0.5 g) was sandwiched between two layers of quartz wool in a quartz tubular microreactor (diameter of 7 mm) placed in an electrical furnace. The increase of the temperature was made in steps of 50°C, from 50 to 550°C. At every predetermined temperature, as a result of catalytic combustion, the gas concentration at the exit of reactor will be smaller than the inlet gas concentration. The catalytic activity of the perovskite catalysts under study was evaluated in terms of the conversion degree of gases over catalysts calculated as [28, 29, 39]:

$$C = \frac{c_{in} - c_{out}}{c_{in}} \times 100(\%), \tag{6}$$

http://dx.doi.org/10.5772/intechopen.77727

where c_{in} and c_{out} are the inlet and outlet gas concentration, respectively, measured by a photo-ionization detector (PID-TECH) for gases and VOCs. Data were collected when the flameless catalytic combustion had reached a steady state, after about 20 min at each temperature. These experiments were repeated decreasing the temperature and similar results were obtained, indicating the stability of perovskites over time, and the absence of the deactivation [39–41].

4.2. Results and discussion

4.2.1. Structural properties

The XRD patterns at room temperature of the analyzed perovskites are shown in **Figure 3**. From the analysis of these diffractograms ensue that samples have a good crystallinity in the specified thermal treatment conditions. All samples were indexed as perovskite structures, without the presence of any foreign phase.

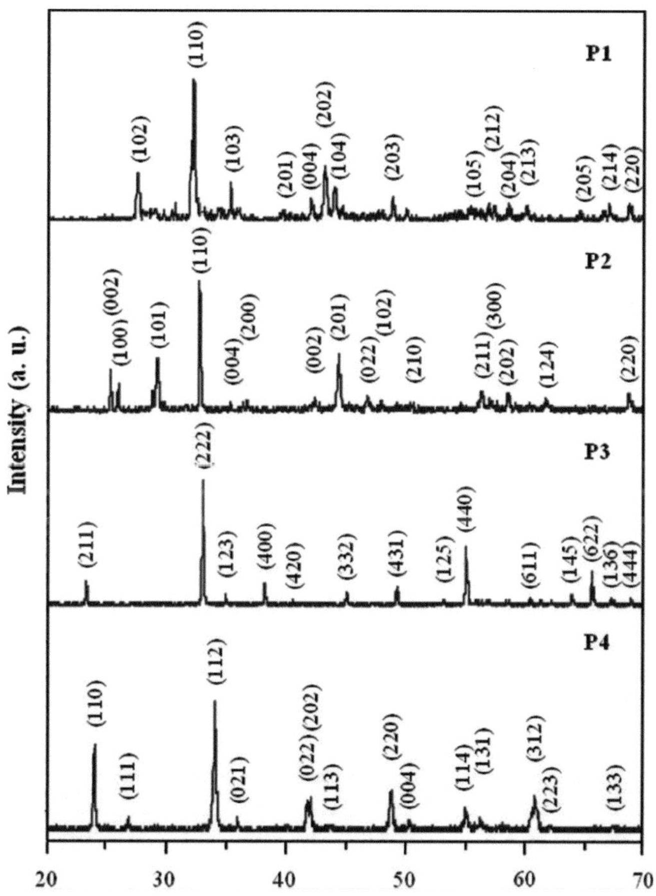

Figure 3. X-ray diffraction patterns of the investigated perovskites.

The structure of crystalline phases, average crystallite size D_{XRD}, derived from XRD data, specific surface area S_{BET} and pore volume obtained from nitrogen adsorption/desorption isotherms at 77 K, are given in **Table 1**.

The average crystallite size derived from XRD data was found to be in the range of 40–89 nm. The smallest crystallites were identified in the $GdAlO_3$ perovskite.

The surface morphology of the studied perovskites is shown in SEM micrographies given in **Figure 4(a–d)**. Generally, the samples are characterized by a very fine structure being composed of aggregates of nanograins with irregular shapes and sizes, with a pronounced intergranular porosity and channels that are favoring the adsorption or desorption of the gas around particle agglomerates.

The chemical elemental composition of the powders heat treated was confirmed by the energy dispersive X-ray spectra (EDX). The obtained chemical elemental composition is typical for this perovskite (any foreign element is absent). **Figure 4(e–h)** presents the EDX spectrum for this perovskites. Moreover, the composition of the sample is similar to the nominal one, ABO_3, that is, the A/(A + B) or B/(A + B) ratio is close to 0.5 (where A and B are given in at.%).

4.2.2. Catalytic activity of perovskites

The catalytic performances of the studied perovskites in the flameless combustion of some VOCs (acetone, propane, benzene and Pb free gasoline) were investigated in the temperature range 50–550°C. The results are presented in **Figure 5(a–d)**, where the gas conversion was plotted as a function of the reaction temperature for each perovskite composition.

The catalysts exhibited substantial differences in catalytic activity, in the studied temperature range. Increasing the reaction temperature facilitates gas combustion. The gas combustion over $SrMnO_3$ catalyst started at much lower temperatures (at about 100°C) compared to the other perovskites. The greater activity of $SrMnO_3$ catalyst toward the conversion of VOCs indicates the availability of reactive oxygen species on the catalyst surface. This suggests oxygen may be less anchored on the perovskite surface and be more available for VOCs oxidation, probably because of the presence of a noteworthy amount of Mn^{4+} ions [39, 41].

The $MnFeO_3$ and $SrCoO_3$ catalysts exhibited high catalytic activity only toward acetone conversion and proved poor catalytic performance in catalytic combustion of propane, benzene and gasoline. This behavior may be related to some rearrangement of their lattice structure and, as a consequence, of the active site structure, on which the catalytic properties of these perovskites depend. It is interesting to point out the strong influence of temperature on the

Sample symbol	Crystalline phases	D_{XRD} (nm)	S_{BET} ($m^2 \cdot g^{-1}$)	Pore volume ($cm^3 \cdot g^{-1}$)
P1	Hexagonal	88.9	2.2	0.0010
P2	Hexagonal	59.9	1.9	0.0030
P3	Cubic	59.2	3.2	0.0044
P4	Orthorhombic	39.6	9.8	0.0018

Table 1. Structure characteristics of investigated perovskites.

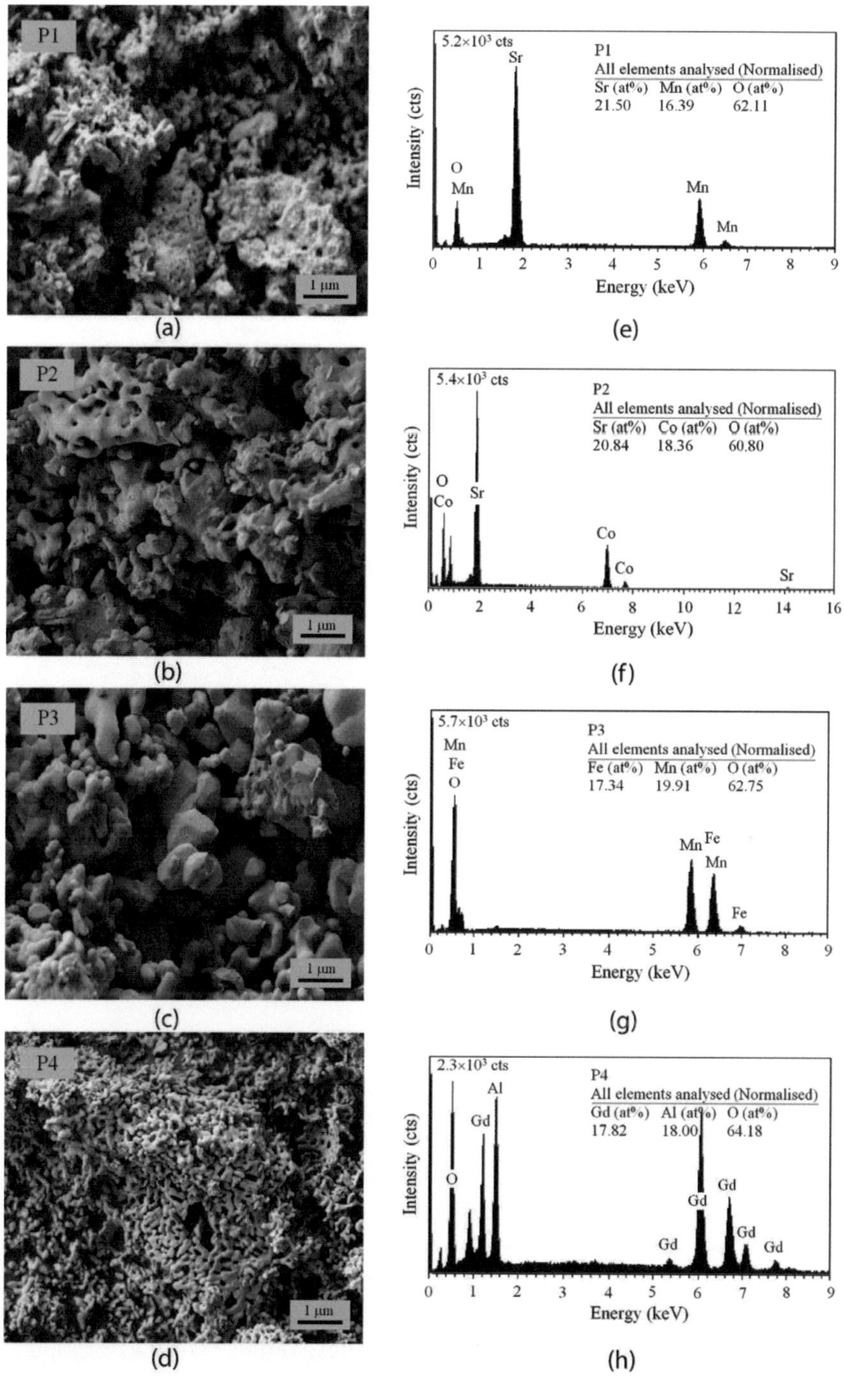

Figure 4. SEM micrographs (a–d) and EDX spectra (e–h) for the studied perovskites.

acetone conversion over the $MnFeO_3$ catalyst. The acetone conversion started at a low temperature (150°C) and the conversion degree sharply increased from 10 to 80% as the temperature increased from 200 to 300°C (**Figure 5(a)**). This behavior was not observed with the

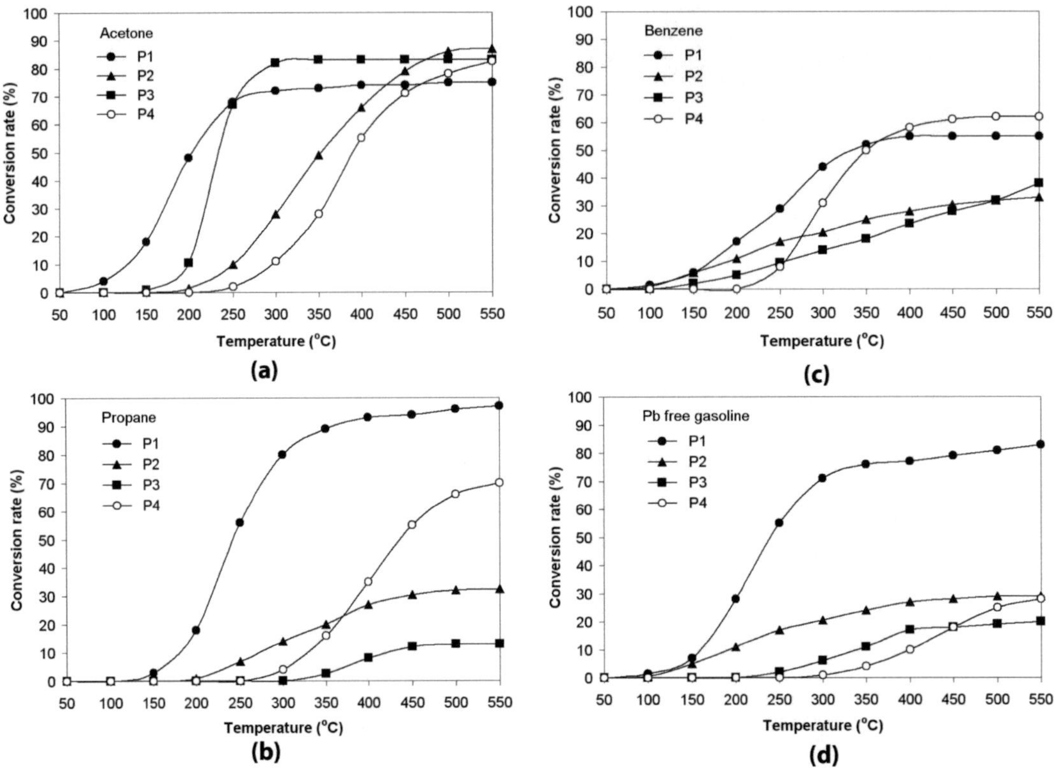

Figure 5. Conversion vs. temperature for catalytic flameless combustion of acetone (a), propane (b), benzene (c) and Pb free gasoline (d).

other perovskites. In contrast with the $MnFeO_3$ catalyst, the $SrCoO_3$ catalyst started the acetone conversion near 200°C and 80% acetone conversion was obtained at a higher temperature, of 450°C [41].

In **Table 2** are given the values of the gas conversion at 500°C and the values of the kinetic parameters (reaction rate and apparent activation energy) for gas oxidation over the studied perovskite catalysts. The apparent activation energies for the catalytic reactions were calculated by means of the Arrhenius type plot of the natural logarithm of the rate constant k at low conversion (below 10–15%) versus inverse temperature (1/T). This plot is a straight line, and from its slope, the apparent activation energy was calculated. One can observe that the reaction rate normalized to specific area changes from 3.8×10^{-2} to 140×10^{-2} μmol s^{-1} m^{-2}. The higher the reaction rate is higher, the more active is the catalyst. Note also the wide variation in activation energies from 40 kJ/mol ($SrCoO_3$ toward gasoline combustion) to 98 kJ/mol ($MnFeO_3$ toward acetone combustion). The smallest values (31–37 kJ/mol) of the apparent activation energy were obtained for the $SrMnO_3$ catalyst with the best catalytic activity.

The differences observed between the activation energies suggest that the nature of the catalytic sites differs from one perovskite to the others. The smaller values of the activation

VOC	Sample	$Conv_{500}$ (%)	Reaction rate ($\mu mol\ s^{-1}\ m^{-2}$)	Activation energy ($KJ\ mol^{-1}$)
Acetone	P1	75	140×10^{-2}	37
	P2	85	16×10^{-2}	41
	P3	85	65×10^{2}	98
	P4	78	67×10^{-2}	89
Propane	P1	95	9.8×10^{-2}	31
	P2	32	3.8×10^{-2}	48
	P3	32	6.1×10^{-2}	80
	P4	65	7.8×10^{-2}	71
Benzene	P1	55	26×10^{-2}	35
	P2	32	56×10^{-2}	44
	P3	32	17×10^{-2}	45
	P4	62	6.5×10^{-2}	68
Pb free gasoline	P1	83	55×10^{-2}	36
	P2	30	11×10^{-2}	40
	P3	20	12×10^{-2}	47
	P4	25	4.2×10^{-1}	62

[a]Reaction rate for VOC concentration at low conversion per unit surface area of catalyst.
[b]Apparent activation energy for low conversions.

Table 2. VOCs conversion at 500°C and kinetic parameters (reaction rate[a] and activation energy[b]) for perovskite catalysts.

energy for the samples $SrCoO_3$ and $SrMnO_3$ can indicate a contribution of the mass transfer effects [41, 42]. The values of the kinetic parameters obtained by us are comparable to those presented by other authors for other perovskite type oxides and VOCs [18, 41, 43].

The $SrCoO_3$ and $MnFeO_3$ catalysts were able to convert 85% acetone at 500°C, whereas the conversion of the other gases was below 30%. The $SrMnO_3$ catalyst was able to convert 95% propane, 83% Pb free gasoline, and 75% acetone at 500°C. The low specific area of the $SrMnO_3$ perovskite does not seem to take part in the higher activity of this catalyst. The improved catalytic activity of $SrMnO_3$ may be ascribed to the higher oxygen mobility due to the oxygen vacancies generated by the presence of manganese ion with variable valence. The difference in the catalytic activity of the studied perovskites cannot be explained by their different specific surfaces. There are many hypotheses about the mechanism of oxidation of VOCs on the oxide compounds. According to one accepted opinion, at low temperatures (below 400°C), the catalytic activity of the perovskite oxides in the total oxidation reactions of the gases is largely determined by the amount of weakly bound surface oxygen species which in turn depends on the presence of oxygen vacancies [41, 44]. The weaker the oxygen binding at the catalyst surface is, the more active the catalyst for complete oxidation of gases is [45–47]. The surface oxygen species (O, O_2^-, O^{2-}) involved in the catalytic combustion may come from the gaseous molecular oxygen and from the lattice oxygen [41, 48, 49]. The interaction of surface active oxygen species with reactants has an important role in mechanism of VOCs total oxidation (suprafacial mechanism) over perovskites oxide, and this is widely accepted to explain the gas oxidation over mixed oxide catalysts [39, 41, 44].

5. Conclusions

The oxide nanostructured materials, especially the perovskites, can have special catalytic combustion properties, and during the last years, there are several works and accomplishments in this connection. Such materials are especially interesting due to their high thermal stability, reduced cost, wide diversity, selectivity to various reactions, as well as to their availability in large amount for wide consumption products (exhaust muffler, filtering installations, addition in liquid combustibles, cigarettes filters, etc.), as well as for work and environment protection equipment.

A recent method of nanoparticles preparation, including those with perovskite structure is the precursor method of self-combustion, which uses as colloidal precipitation medium the polyvinyl alcohol. This method permits to obtain oxide compounds with homogeneous nanograined porous highly pure structure and reproducible properties. It is superior to other known conventional methods.

By the precursor method of self-combustion followed by heat treatment in air, several nanostructured perovskites with various compositions ($SrMnO_3$, $SrCoO_3$, $MnFeO_3$ and $GdAlO_3$) were prepared for catalyst applications. The catalytic tests of the perovskites in the flameless catalytic combustion of acetone, propane, benzene, and Pb free gasoline evidenced that the degree of the catalytic activity varied considerably with the composition of perovskite. The $SrMnO_3$ catalyst is generally more active than the other catalysts. The $SrMnO_3$ catalyst is able to convert 95% propane, 83% Pb free gasoline and 75% acetone at 500°C. The $SrCoO_3$ and $MnFeO_3$ catalysts are able to convert 85% acetone at 500°C whereas the conversion of the other gases was below 30%. It is worth noting the sharp increase of the acetone conversion over $MnFeO_3$, from 10 to 80% as the temperature increases from 200 to 300°C. The difference in the catalytic activity of the studied perovskites cannot be explained by their different specific surfaces. A possible explanation for the change in the catalytic properties as result of modification of perovskite composition may be either the different reactivity of the active oxygen species involved in the catalytic oxidation, or the variation in the number of active sites on the perovskite surface determined by the specific structural properties of each perovskite [41, 50].

Conflict of interest

The author has declared no conflict of interest.

Author details

Corneliu Doroftei

Address all correspondence to: docorneliug@gmail.com

Integrated Center for Studies in Environmental Science for North-East Region, Alexandru Ioan Cuza University of Iasi, Iasi, Romania

References

[1] Mellado JD, Kindelan M, Sanchez AL. A simplified formulation for heterogeneous catalytic combustion in stagnation-point flows. Combustion and Flame. 2003;**132**:596-599. DOI: 10.1016/S0010-2180(02)00493-5

[2] Masatake H. When gold is not noble: Catalysis by nanoparticles. The Chemical Record. 2003;**3**:75-87

[3] Suzuki Y, Horii Y, Kasagi N. Microcatalytic combustor with tailored Pt/Al_2O_3 films, In: Proceedings of the 3rd International Workshop on Micro and Nanotechnology for Power Generation and Energy Conversion Applications (Power MEMS 2003); 4-5 December 2003; Makuhari; Japan. pp. 23-26

[4] Mongkhonsi T, Umpo S, Thongsang P. Oxidation of anhydrides over MgO promoted cobalt and vanadium oxide catalysts. Chemistry Letters. 2002;**31**:978-979. DOI: 10.1246/cl.2002.978

[5] Cimino S, Colonna S, De Rossi S, Faticanti M, Lisi L, Pettiti I, Porta P. Methane combustion and CO oxidation on zirconia-supported La, Mn oxides and $LaMnO_3$ perovskite. Journal of Catalysis. 2002;**205**:309-317. DOI: 10.1006/jcat.2001.3441

[6] Suhadi DR, Awang M, Hassan MN, Abdullah R, Azizi H. Review of photochemical smog pollution in Jakarta Metropolitan, Indonesia. American Journal of Environmental Sciences. 2005;**1**:110-118

[7] Zhang R, Alamdari H, Bassir M, Kaliaguine S. Optimization of mixed Ag catalysts for catalytic conversions of NO and C_3H_6. International Journal of Chemical Reactor Engineering. 2007;**5** Article A70

[8] Andreeva D, Petrova P, Ilieva L, Sobczak JW, Abrashev MV. Design of new gold catalysts supported on mechanochemically activated ceria-alumina, promoted by molybdena for complete benzene oxidation. Applied Catalysis B: Environmental. 2008;**77**:364-372. DOI: 10.1016/j.apcatb.2007.08.009

[9] Pecchi G, Reyes P, Figueroa A, Fierro JLG. Catalytic combustion on toluene on $Pd\text{-}Cu/SiO_2$ catalysts. Boletin de la Sociedad Chilena de Quimica. June 2000;**4**(2):1-7. DOI: 10.4067/S0366-16442000000200007

[10] Moreira ML, Paris EC, do Nascimento GS, Longo VM, Sambrano JR, Mastelaro VR, Bernardi MIB, Andre's J, Varela JA, Longo E. Structural and optical properties of $CaTiO_3$ perovskite-based materials obtained by microwave-assisted hydrothermal synthesis: An experimental and theoretical insight. Acta Materialia. 2009;**57**:5174-5185. DOI: 10.1016/j.actamat.2009.07.019

[11] Costa ACFM, Diniz APA, Gama L, Morelli MR, Kiminami RHGA. Comparison of Ni-Zn ferrite powder preparation by combustion reaction using different syntesization routes. Journal of Metastable and Nanocrystalline Materials. 2004;**20-21**:582-587. DOI: 10.4028/www.scientific.net/JMNM.20-21.582

[12] Svensson EE, Nassos S, Boutonnet M, Jaras SG. Microemulsion synthesis of MgO-supported $LaMnO_3$ for catalytic combustion of methane. Catalysis Today. 2006;**117**:484-490. DOI: 10.1016/j.cattod.2006.06.014

[13] Zhang X, Liu Z, Wei Z, Du X, Gong M, Chen Y. Preparation of high performance methane combustion catalyst and its application to natural gas catalytic combustion fan-boiler. Chinese Journal of Catalysis. 2006;**27**:823-826. DOI: 10.1016/S1872-2067(06)60045-4

[14] Xanthopoulou G. Catalytic properties of the SHS products—Review. Advances in Science and Technology. 2010;**63**:287-296. DOI: 10.4028/www.scientific.net/AST.63.287

[15] Battiston AA, Bitter JH, Heijboer WM, de Groot FMF, Koningsberger DC. Reactivity of Fe-binuclear complexes in over-exchanged Fe/ZSM5 studied by in situ XAFS spectroscopy. Journal of Catalysis. 2003;**215**:279-293. DOI: 10.1016/S0021-9517(02)00188-4

[16] Resini C, Catania F, Berardinelli S, Paladino O, Busca G. Catalytic wet oxidation of phenol over lanthanum strontium manganite. Applied Catalysis B: Environmental. 2008;**84**:678-683. DOI: 10.1016/j.apcatb.2008.06.005

[17] Hirabayashi D, Yoshikawa T, Kawamoto Y, Mochizuki K, Suzuki K. Characterization and application of calcium ferrites based materials containing active oxygen species. Advances in Science and Technology. 2006;**45**:2169-2175

[18] Milt VG, Spretz R, Ulla MA, Lombardo EA, Garcia Fierro JL. The nature of active sites for the oxidation of methane on La-based perovskites. Catalysis Letters. 1996;**42**:57-63. DOI: 10.1007/BF00814467

[19] Hammami R, Aissa SB, Batis H. Effects of thermal treatment on physico-chemical and catalytic properties of lanthanum manganite $LaMnO_{3+y}$. Applied Catalysis A: General. 2009;**353**:145-153. DOI: 10.1016/j.apcata.2008.10.048

[20] Perez-Ramírez J, Vigeland B. Lanthanum ferrite membranes in ammonia oxidation Opportunities for 'pocket-sized' nitric acid plants. Catalysis Today. 2005;**105**:436-442. DOI: 10.1016/j.cattod.2005.06.057

[21] Batis NH, Delichere P, Batis H. Physicochemical and catalytic properties in methane combustion of $La_{1-x}Ca_xMnO_{3\pm y}$ ($0 \le x \le 1$; $-0.04 \le y \le 0.24$) perovskite-type oxide. Applied Catalysis A: General. 2005;**282**:173-180. DOI: 10.1016/j.apcata.2004.12.009

[22] Ponce S, Pena MA, Fierro JLG. Surface properties and catalytic performance in methane combustion of Sr-substituted lanthanum manganites. Applied Catalysis B. 2000;**24**:193-205. DOI: 10.1016/S0926-3373(99)00111-3

[23] Tomatis M, Xu HH, He J, Zhang XD. Recent development of catalysts for removal of volatile organic compounds in flue gas by combustion: A Review. Journal of Chemistry. 2016;Article ID 8324826::1-15. DOI: 10.1155/2016/8324826

[24] Alifanti M, Florea M, Pârvulescu VI. Ceria-based oxides as supports for $LaCoO_3$ perovskite; catalysts for total oxidation of VOC. Applied Catalysis B. 2007;**70**:400-405. DOI: 10.1016/j.apcatb.2005.10.037

[25] Zhang R, Luo N, Yang W, Liu N, Chen B. Low-temperature selective catalytic reduction of NO with NH_3 using perovskite-type oxides as the novel catalysts. Journal of Molecular Catalysis A: Chemical. 2013;**731**:86-93. DOI: 10.1016/j.molcata.2013.01.018

[26] Huang HF, Sun Z, Lu HF, Shen LQ, Chen YF. Study on the poisoning tolerance and stability of perovskite catalysts for catalytic combustion of volatile organic compounds. Reaction Kinetics Mechanisms and Catalysis. 2010;**101**:417-427. DOI: 10.1007/s11144-010-0235-6

[27] Doroftei C, Popa PD, Rezlescu E, Rezlescu N. Structural and catalytic characterization of nanostructured iron manganite. Composites: Part B. 2014;**67**:179-182. DOI: 10.1016/j.compositesb.2014.07.005

[28] Doroftei C, Popa PD, Rezlescu E, Rezlescu N. Nanocrystalline $SrMnO_3$ powder as catalyst for hydrocarbon combustion. Journal of Alloys and Compounds. 2014;**584**:195-198. DOI: 10.1016/j.jallcom.2013.09.054

[29] Leontie L, Doroftei C. Nanostructured spinel ferrites for catalytic combustion of gasoline vapors. Catalysis Letters. 2017;**147**:2542-2548. DOI: 10.1007/s10562-017-2164-8

[30] Rezlescu N, Rezlescu E, Popa PD, Doroftei C, Ignat M. Nanostructured $GdAlO_3$ perovskite, a new possible catalyst for combustion of volatile organic compounds. Journal of Materials Science. 2013;**48**:4297-4304. DOI: 10.1007/s10853-013-7243-7

[31] Popa PD, Doroftei C, Rezlescu E, Rezlescu N. Influence of colloidal environment on the properties of nanoferrites MgCuZn prepared by precursor method of self-combustion. Journal of Optoelectronics and Advanced Materials. 2010;**12**:876-880

[32] Rezlescu E, Doroftei C, Popa PD, Rezlescu N. The transport and magnetic properties of La-Pb-Mg-Mn-O manganites at low temperatures. Journal of Magnetism and Magnetic Materials. 2008;**320**:796-802. DOI: 10.1016/j.jmmm.2007.08.019

[33] Doroftei C, Leontie L, Popa A. The study on nanogranular system manganites La-Pb-Ca-Mn-O which exhibits a large magnetoresistance near room temperature. Journal of Materials Science-Materials in Electronics. 2017;**28**:12891-12899. DOI: 10.1007/s10854-017-7119-8

[34] Rezlescu N, Doroftei C, Rezlescu E, Popa PD. The influence of Sn^{4+} and/or Mo^{6+} ions on the structure, electrical and gas sensing properties of Mg-ferrite. Physica Status Solidi A. 2006;**203**:306-316. DOI: 10.1002/pssa.200521043

[35] Rezlescu N, Iftimie N, Rezlescu E, Doroftei C, Popa PD. Semiconducting gas sensor for acetone based on the fine grained nickel ferrite. Sensors and Actuators B. 2006;**114**:427-432. DOI: 10.1016/j.snb.2005.05.030

[36] Doroftei C, Rezlescu E, Rezlescu N, Popa PD. Gas sensing properties of samarium substituted lithium ferrite. Journal of Optoelectronics and Advanced Materials. 2008;**10**:2390-2395

[37] Doroftei C. Formaldehyde sensitive Zn-doped LPFO thin films obtained by rf sputtering. Sensors and Actuators B-Chemical. 2016;**231**:793-799. DOI: 10.1016/j.snb.2016.03.104

[38] Doroftei C, Popa PD, Iacomi F. Selectivity between methanol and ethanol gas of La-Pb-Fe-O perovskite synthesized by novel method. Sensors and Actuators A-Physical. 2013;**190**:176-180. DOI: 10.1016/j.sna.2012.11.018

[39] Doroftei C, Leontie L. Synthesis and characterization of some Nanostructured composite oxides for low temperature catalytic combustion of dilute propane. RCS Advances. 2017;**7**:27863-27871. DOI: 10.1039/c7ra03916f

[40] Rezlescu N, Rezlescu E, Popa PD, Doroftei C, Ignat M. Some nanograined ferrites and perovskites for catalytic combustion of acetone at low temperature. Ceramics International. 2015;**41**:4430-4437. DOI: 10.1016/j.ceramint.2014.11.134

[41] Rezlescu N, Rezlescu E, Popa PD, Doroftei C, Ignat M. Characterization and catalytic properties of some perovskites. Composites: Part B. 2014;**60**:515-522. DOI: 10.1016/j.compositesb.2014.01.006

[42] Florea M, Alifanti M, Parvulescu VI, Mihaila-Tarabasanu D, Diamandescu L, Feder M, Negrila C, Frunza L. Total oxidation of toluene on ferrite-type catalysts. Catalysis Today. 2009;**141**:361-366. DOI: 10.1016/j.cattod.2008.05.005

[43] Song KS, Klvana D, Kirchnerova J. Kinetics of propane combustion over $La_{0.66}Sr_{0.34}Ni_{0.3}Co_{0.7}O_3$ perovskite. Applied Catalysis A. 2001;**213**:113-121

[44] Ivanov DV, Pinaeva LG, Sadovskaya EM, Isupova LA. Influence of the mobility of oxygen on the reactivity of $La_{1-x}Sr_{(x)}MnO_3$ perovskites in methane oxidation. Kinetics and Catalysis. 2011;**52**:401-408. DOI: 10.1134/S0023158411030086

[45] Yakovleva IS, Isupova IA, Rogov VA, Sadykov VA. Forms of oxygen in $La_{1-x}Ca_xMnO_{3+delta}$ (x = 0-1) perovskites and their reactivities in oxidation reactions. Kinetics and Catalysis. 2008;**49**:261-270. DOI: 10.1134/S0023158408020146

[46] Isupova LA, Yakovleva IS, Alikina GM, Rogov VA, Sadykov VA. Reactivity of $La_{(1-x)}Sr_{(x)}FeO_{(3\_y)}$ (x = 0-1) perovskites in oxidation reactions. Kinetics and Catalysis. 2005;**46**:729-735. DOI: 10.1007/s10975-005-0129-1

[47] Marchetti L, Forni L. Catalytic combustion of methane over perovskites. Applied Catalysis B. 1998;**15**:179-187. DOI: 10.1016/S0926-3373(97)00045-3

[48] Li WB, Wang JX, Gong H. Catalytic combustion of VOCs on non-noble metal catalysts. Catalysis Today. 2009;**148**:81-87. DOI: 10.1016/j.cattod.2009.03.007

[49] Kundakovic L, Flytzani-Stefanopoulos M. Cu- and Ag-modified cerium oxide catalysts for methane oxidation. Journal of Catalysis. 1998;**179**:203-221. DOI: 10.1006/jcat.1998.2213

[50] Sugahara K, Kamata K, Muratsugu S, Hara M. Amino acid-aided synthesis of a hexagonal $SrMnO_3$ nanoperovskite catalyst for aerobic oxidation. ACS Omega. 2017;**2**:1608-1616. DOI: 10.1021/acsomega.7b00146

Chapter 6

Proton Exchange Membrane Water Electrolysis as a Promising Technology for Hydrogen Production and Energy Storage

Radenka Maric and Haoran Yu

Additional information is available at the end of the chapter

http://dx.doi.org/10.5772/intechopen.78339

Abstract

Proton exchange membrane (PEM) electrolysis is industrially important as a green source of high-purity hydrogen, for chemical applications as well as energy storage. Energy capture as hydrogen via water electrolysis has been gaining tremendous interest in Europe and other parts of the world because of the higher renewable penetration on their energy grid. Hydrogen is an appealing storage medium for excess renewable energy because once stored, it can be used in a variety of applications including power generation in periods of increased demand, supplementation of the natural gas grid for increased efficiency, vehicle fueling, or use as a high-value chemical feedstock for green generation of fertilizer and other chemicals. Today, most of the cost and energy use in PEM electrolyzer manufacturing is contributed by the cell stack manufacturing processes. Current state-of-the-art electrolysis technology involves two options: liquid electrolyte and ion exchange membranes. Membrane-based systems overcome many of the disadvantages of alkaline liquid systems, because the carrier fluid is deionized water, and the membrane-based cell design enables differential pressure operation.

Keywords: hydrogen, renewable energy, proton exchange membrane (PEM) water electrolysis, nanostructure, oxygen evolution reaction (OER)

1. Hydrogen as a future energy carrier

Eighty-five percent of the energy consumed globally is provided by fossil fuels, namely coal, oil, and natural gas [1]. Fossil fuels come from finite resources which will eventually become scarce and difficult to explore. Thus, fossil fuels are considered nonrenewable energy sources [2]. Furthermore, consuming fossil fuels produces greenhouse gases and other byproducts, causing climate change and air pollution. The growing demand for energy requires a rapid shift from fossil fuels to renewable energy sources, such as wind, solar, biomass, hydropower, and geothermal energy [3]. In this context, hydrogen was proposed as a promising candidate for a secondary source of energy as early as 1973 [4]. Being a potential energy carrier in the future, hydrogen plays an important role in the path toward a low-carbon energy structure that is environmentally friendly [5–10].

1.1. Hydrogen production from renewable energy sources

Currently, the steam reforming process is the most economical way of producing hydrogen. In fact, as much as 96% of hydrogen is made from hydrocarbon fuels [5], which neither address the dependence on finite resources nor reduce the amount of carbon from the energy structure. An alternative way of producing hydrogen is the power-to-gas strategy where intermittent energy resources are transferred and stored as hydrogen (**Figure 1**). Here, hydrogen is mainly produced from water electrolysis where water is split into hydrogen and oxygen by supplying electrical energy:

$$H_2O \xrightarrow{electricity} H_2 + \frac{1}{2}O_2 \tag{1}$$

In an electrolyzer, the above reaction is separated by an electrolyte (either in liquid or solid form) into two half reactions. The hydrogen evolution reaction (HER) occurs at the cathode:

$$2H^+ + 2e^- \rightarrow H_2\uparrow, E^0 = 0.0\ V \tag{2}$$

While the oxygen evolution reaction (OER) occurs at the anode:

$$H_2O \rightarrow 2H^+ + 2e^- + \frac{1}{2}O_2\uparrow, E^0 = 1.23\ V \tag{3}$$

Water electrolysis technologies are classified into three categories based on the applied electrolyte: alkaline water electrolysis, proton exchange membrane (PEM) water electrolysis, and solid oxide water electrolysis [11]. PEM water electrolysis systems provide several advantages over the other two electrolysis technologies, such as higher rate of hydrogen production, more compact design, and greater energy efficiency [12–15]. Compared to alkaline electrolysis, the solid electrolyte membrane in PEM electrolysis reduces the hydrogen crossover significantly and thus allows for high-pressure operation. In addition, as required by the role of electrolytic hydrogen production in renewable energy storage, dynamic response of PEM water electrolysis is superior to alkaline electrolysis or solid oxide electrolysis. The large quantity of liquid electrolyte in alkaline electrolysis requires the proper temperature to be maintained and could raise issues for a cold start. On the other hand, solid oxide electrolysis operated in a temperature range of 500–700°C is more suitable for constant operation than a dynamic response where the heat-up step could be slow.

The produced hydrogen can have several pathways to different applications (**Figure 1**). It can be utilized for hydrogen fueling stations to power fuel cell vehicles or feed the combined heat and power (CHP) units for household uses. Moreover, the electrolytic hydrogen can be used as

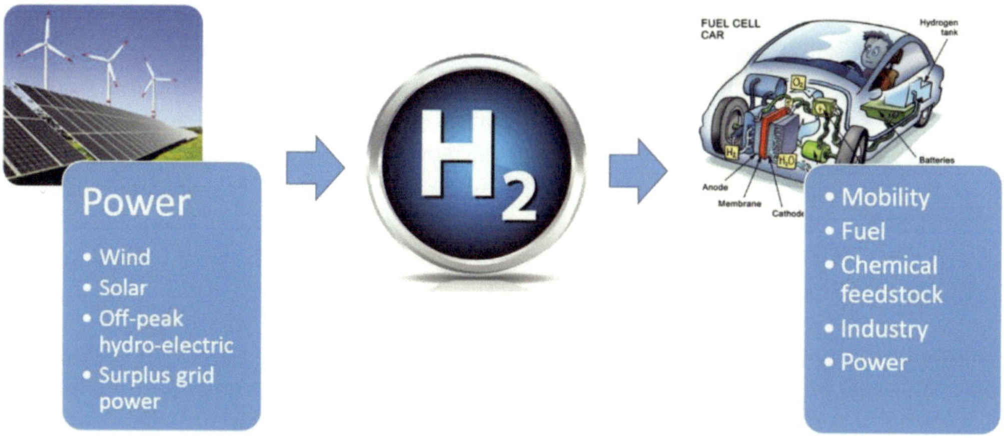

Figure 1. Schematic diagram of power-to-gas strategy.

chemical feedstock in methanation after combining with CO_2 stream from biogas or flue gas to produce renewable natural gas. Further, the generated hydrogen can also be consumed as a raw material by hydrogen users such as oil refining and semiconductor industry. Finally, the hydrogen can be transferred to electricity when the grid demand is high.

Hydrogen can also be produced from biomass via pyrolysis or gasification. Wood, agricultural crops and its byproducts, organic waste, animal waste, waste from food processing, and so on are all sources of biomass. Biomass pyrolysis is basically [9]:

$$Biomass + Heat \xrightarrow{625-775K,\,1-5\ bar} CH_4 + CO + H_2 + other\ products \quad (4)$$

Employing catalysts, such as Ni-based catalysts, can enhance the yield of hydrogen from biomass pyrolysis. Moreover, hydrogen production can be improved by introducing steam reforming and water-gas shift reaction to the pyrolysis [9]. For the gasification process, biomass is pyrolyzed at higher temperatures producing mostly gaseous products [9]:

$$Biomass + O_2\ or\ H_2O \xrightarrow{875-1275K} CH_4, CO, H_2, CO_2, H_2O + other\ hydrocarbons + char + tar + ash \quad (5)$$

It is beneficial that biomass pyrolysis and gasification can be operated in small scale and at remote locations, which reduces the cost of hydrogen transportation and storage and improves the availability of hydrogen to end consumers [16]. In addition, pyrolysis and gasification can consume a wide range of biomass feedstocks [16]. Therefore, biomass is recognized as a major renewable and sustainable energy source to replace fossil fuel.

1.2. Hydrogen storage

Currently, hydrogen storage in high-pressure vessel is the most widely used method [7]. However, hydrogen is pressurized up to 700 bar for practical purposes such as the refueling time at a hydrogen station or the driving range for a fuel cell vehicle [17]. Hydrogen compression to 700 bar consumes a lot of energy that makes the volumetric energy density decrease from 10 to 5.6 MJ/L, much lower than gasoline (34 MJ/L) [6, 17]. Therefore, solid-state storage is usually coupled with high-pressure hydrogen vessels. For example, hydrogen can be stored in the interstitial sites of metal hydride crystals [17]. This method achieves higher volumetric energy density at room temperature than liquid hydrogen and consumes less operating energy for storage. Thus, metal hydride cartridge is suitable for portable application due to the convenience of refill/replace [7, 18]. In addition, with appropriate hydrogen refill and release properties at room temperature, metallic hydrides are good for stationary energy storage [17]. One drawback of solid-state storage is that metallic hydrides contain heavy transition metals, which reduce the gravimetric energy density of the device [19].

In summary, hydrogen offers several advantages as an energy carrier: its combustion produces energy and only water that is carbon-free as a byproduct; it can be produced from renewable and sustainable sources; its energy can be distributed quite easily, in accordance with the end user's requests and with the development of new technologies for transportation and storage; it may be used in both centralized or distributed energy production [7]. In spite of these advantages, hydrogen has failed to be widely used in energy systems due to numerous barriers, including costs of production and storage and the availability of infrastructure [10]. This chapter focuses on the technological challenges of PEM water electrolyzers for hydrogen production.

2. Introduction of PEM water electrolysis

2.1. Fundamentals of PEM water electrolysis

Figure 2 shows the schematic diagram of PEM water electrolysis. The electrolysis process is an endothermic process and electricity is applied as the energy source. The water electrolysis reaction is thermodynamically possible at potentials higher than 1.23 V vs. RHE (reversible hydrogen electrode). The thermoneutral potential at which the cell can operate adiabatically is 1.48 V vs. RHE. Typical PEM water electrolysis devices operate at potential well over 1.48 V vs. RHE and heat is generated by the reaction [15]. The PEM water electrolysis system, similar to proton exchange membrane fuel cell (PEMFC), anode and cathode is separated by a solid polymer electrolyte (Nafion) of thickness below 0.2 mm. At the anode, water is oxidized to produce oxygen, electrons, and protons. The protons are transported across the electrolyte membrane to be reduced to hydrogen. The catalyst for water oxidation or oxygen evolution is typically iridium, which can withstand the corrosive environment due to high overpotential on the anode. Water is channeled to the anode by a titanium flow field, and a piece of porous titanium mesh is placed between the anode catalyst layer and the water channel serving as the diffusion layer. The cathode configuration is similar to the PEMFC with Pt-based catalyst and a graphite flow field to transport hydrogen. A piece of carbon paper is used as the gas diffusion layer (GDL) placed between the cathode catalyst and the flow field.

The hydrogen production rate of ideal electrolysis is proportional to the charge transferred, according to Faraday's law. It can be expressed as [20]:

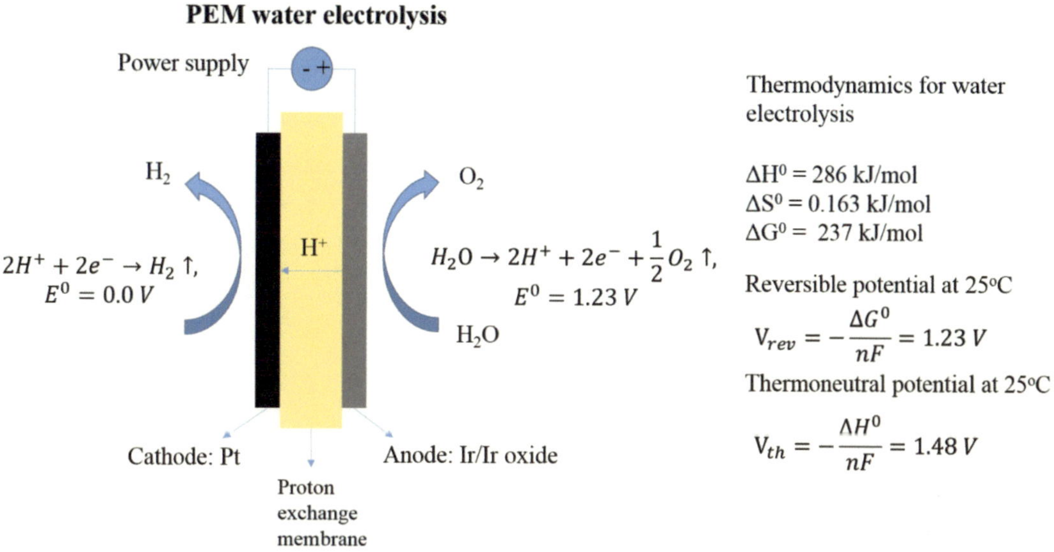

Figure 2. Schematic diagram of PEM water electrolysis and the fundamental thermodynamic properties.

$$f_{H_2} = \eta_F \frac{N_{Cell} I_{Cell}}{zF} \frac{22.41}{1000} 3600 \tag{6}$$

where N_{cell} is the total number of cells in the system and I_{cell} is the electric current. η_F is the Faraday efficiency, or current efficiency, and is defined as the ratio of ideal electric charge and the practical charge consumed by the device when a certain amount of hydrogen is generated. η_F is usually about 0.95 [20]. The specific energy consumption E (kWh/Nm^3) for a given time interval Δt is:

$$\mathrm{E} = \frac{\int_0^{\Delta t} N_{cell} I_{cell} V_{cell} dt}{\int_0^{\Delta t} f_{H2} dt} \tag{7}$$

Another important parameter for PEM water electrolysis is the efficiency:

$$\eta = \frac{HHV}{E} \tag{8}$$

where HHV is the higher heating value of hydrogen (39.4 kWh/kg at STP). Since PEM water electrolysis is usually supplied by liquid water, the HHV assumes that all the heat from the water is recovered by restoring the water temperature to the initial ambient state [20].

2.2. Challenges for PEM water electrolysis: the OER catalyst

Significant advances are needed in catalyst and membrane materials as well as the labor-intensive manufacturing process for PEM water electrolysis to be cost-effective for wide-spread application in renewable energy systems [21]. The state-of-the-art anode catalyst in

References	Anode catalyst	Ir loading, mg cm^{-2}	Cathode catalyst	Pt loading, mg cm^{-2}	Electrode fabrication process	Active area	Test period, hours	Operating temperature, °C	Operating current, A cm^{-2}	Test period, hours	Degradation rate, μV h^{-1}
Grigoriev et al. [40]	Ir black	1.5	Pt/Vulcan XC-72	1.0	Spraying	25	4000	60	0.5	4000	35.5
Rozain et al. [41]	IrO_2/Ti	0.12	Pt/C, TKK	0.25	Spraying	25	1000	80	1	1000	27
Rozain et al. [41]	IrO_2	0.32/0.1	Pt/C, TKK	0.25	Spraying	25	1000	80	1	1000	110/180*
Siracusano et al. [42]	IrO_2	0.4	Pt/Vulcan XC-72	0.1	Spray-coating	5	1000	80	1.0	1000	12
Rakousky et al. [43]	IrO_2 and TiO_2	2.25	Pt/C	0.8	Commercial	17.64	1150	80	2.0	1150	194
Lettenmeier et al. [44]	Ir black	1	Pt black	0.9	Commercial	120	400	55–60	2.0	400	Not significant
Lewinski et al. [45]	Ir-NSTF	0.25	Pt-NSTF	0.25	3 M	50	5000	80	2	5000	Average 6.8
RSDT Cell-4	IrOx and Nafion	0.08	Pt/Vulcan XC-72R	0.3	RSDT	86	4543	80	1.8	4543	36.5–48.7, 11.5
Wang et al. [46]	$Ir_{0.7}Ru_{0.3}Ox$	1**	Pt/C	0.4	Air-brush spraying	25	400	80	1	400	Not significant
Siracusano et al. [47]	$Ir_{0.7}Ru_{0.3}Ox$	0.34**	Pt/Vulcan XC-72	0.1	Electrode fabrication process	Active area	1.0/3.0	80	1.0/ 3.0	1000	15/23***

*Degradation rate at 0.32 and 0.1 mg cm^{-2} Ir loading, respectively.
**Loading of Ir and Ru combined.
***Degradation rate at 1.0 and 3.0 A cm^{-2}, respectively.

Table 1. Reported degradation rates for iridium-based catalysts in long-term electrolyzer tests.

conventional PEMWEs is iridium oxide (IrOx) or mixed oxide with ruthenium [22, 23]. Typical catalysts for commercial electrodes have IrOx loading from 1 to 3 mg cm^{-2} [24]. This level of catalyst loading is too high to meet the long-term cost targets for energy markets [23, 25, 26]. Furthermore, while using current electrolysis technology, the translation of catalyst development from lab scale to the megawatt scale remains challenging in terms of catalyst cost and stability [25].

For Ir- and Ru-based OER catalyst, a balance between activity and stability has been reported [27, 28], which strongly depends on the chemical structure and surface properties of the oxide [23]. The oxide properties, however, are highly sensitive to synthesis conditions. For Ir-based OER catalyst, most synthesis methods involve a calcination step to improve the catalyst stability [29–35]; however, for this approach, a sacrifice of the OER activity is usually inevitable. Indeed, thermally prepared IrOx shows higher stability but lower activity compared to electrochemically prepared oxide, or hydrous IrOx [30, 31, 35]. The latter is also frequently referred to as amorphous IrOx due to the presence of lower valence Ir (III) oxide [36–38]. According to reference [39], Ir (III) is the major intermediate species for iridium dissolution. Combining both activity and stability remains a challenge for Ir-based catalysts, and it is the major hurdle that limits the reduction of the anode iridium loading in PEM water electrolysis.

The enhancement of catalyst stability is of equal importance as the reduction of catalyst loading. Long-term operation at high current density up to thousands of hours is particularly challenging with an Ir loading less than 1 mg cm^{-2}. Recent literature has seen a growing number of studies on longer-term electrolysis operation ranging from hundreds to several thousand hours [40–47] (**Table 1**). In particular, the best reported stability was on a nanostructured thin film (NSTF) cell, which achieved 5000 hours with a constant current load of 2 A cm^{-2} and 0.25 mg cm^{-2} Ir loading [45]. IrOx supported on Ti catalyst with 50 wt% Ir and a low catalyst loading of 0.12 mg cm^{-2} achieved more than 1000 hours operation at a lower current load of 1 A cm^{-2} [41]. No supported catalyst, other than titanium supported catalyst has been able to achieve such high catalyst stability. [48]

3. The role of nanostructures in PEM electrolyzer performance

3.1. The structure of iridium oxide (IrOx)

The OER performance of IrOx strongly depends on the chemical structure and surface properties of the oxide [23]. Before discussing the influence of oxide structure on the electrolyzer performance, a thorough understanding of the OER mechanism is needed. The OER on IrOx consists of three steps forming a closed circle (**Figure 3**) [39, 49, 50]:

$$HIrO_2 \rightarrow IrO_2 + H^+ + e^- \tag{9}$$

$$IrO_2 + H_2O \rightarrow IrO_2(OH) + H^+ + e^- \tag{10}$$

$$2IrO_2(OH) \rightarrow 2HIrO_2 + O_2. \tag{11}$$

Reaction (9) is fast and in pre-equilibrium. Reaction (10) is the rate-determining step (RDS), and the theoretical Tafel slope of this mechanism is 2RT/3F = 40 mV/dec. Since it is noted that Eq. (11) involves two iridium sites, a binuclear mechanism of OER on IrOx is proposed for Eq. (11) [51]. With the aid of density functional theory (DFT) calculations, Steesgra et al. [52] show that the first step of the binuclear mechanism is the oxidation of Ir (IV) – Ir (IV) to Ir (IV) – Ir (V), which is Eq. (10). Therefore, $IrO_2{\cdot}H_2O$ is considered to be the precursor for the binuclear mechanism. The coverage of Ir (V) increases with the anodic potential until the conditions for the binuclear mechanism are satisfied [39] at the onset of OER.

In a series of studies of iridium dissolution [31, 32, 39, 53], it is proposed that the precursor for dissolution are the oxygenated Ir (III) species (**Figure 3**). The formation and reduction of higher valence Ir (IV) species pass through an Ir (III) intermediate, which is shared in both OER and iridium dissolution. It is concluded in [39] that iridium dissolution has no direct link with OER activity and that they are two pathways sharing an intermediate species, Ir (III). Thus, it is possible to suppress one without alternating the other [39]. Therefore, the ratio of Ir (IV) to Ir (III) on the IrOx surface is a crucial parameter that controls the iridium dissolution and the OER.

Amorphous and nano-sized IrOxs have received much attention as promising OER catalysts due to the co-existence of Ir (III) and Ir (IV) oxidation state [35–38, 54–56], forming an oxide termed "iridium oxohydroxide." For example, using X-ray absorption spectroscopy (XAS) and X-ray absorption near edge spectroscopy (XANES), Minguzzi et al. [37, 38] confirms the co-existence of Ir (III) and Ir (IV) at anodic potentials where OER occurs. However, this does not imply that a mixed IrOx state fundamentally leads to high OER activity. In fact, the OER mechanism (Eqs. (9)–(11)) predicts that Ir (III) and Ir (IV) would certainly co-exist as Eqs. (9)–(11) form a loop. The higher OER activity of mixed Ir (III) and Ir (IV) oxides have been attributed to the presence of electrophilic O^{I-} species [57–61]. Electrophilic O^{I-} is the precursor for the O-O bond formation according to the binuclear mechanism [52]. The oxygen 2p state probed with near-edge X-ray absorption fine structure (NEXAFS) suggests that electrophilic O^{I-} has a lower energy state compared to the O^{II-} species [61]. The weakly bonded O^{I-} is susceptible to nucleophilic attack from the pre-adsorbed water or hydroxyl groups, promoting the formation of O-O

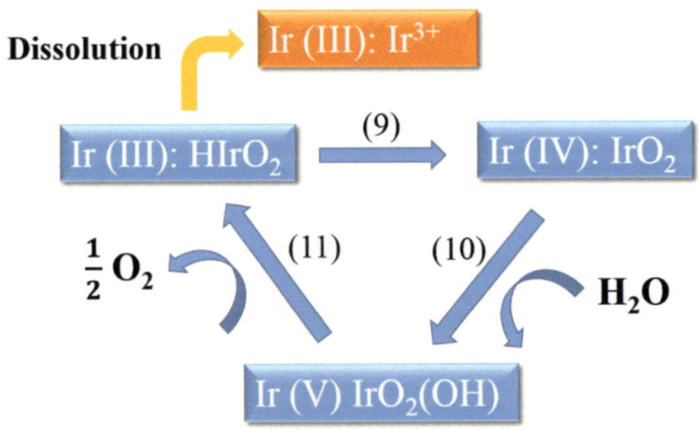

Figure 3. Circle of oxygen evolution: Eqs. (9)–(11).

bond [58, 59]. Calcination at a high temperature of "dry" IrOx (O^{II-} species) where electrophilic O^{I-} becomes strongly bonded O^{II-} species and thus the OER activity is reduced [36], in agreement with other studies [30, 31, 35, 62]. However, the "dry" IrOx is more stable against dissolution as the Ir (III) species on the surface is reduced [32, 60].

In addition, the IrOx structure can be modified when a metal oxide support is incorporated with IrOx nanoparticles via the metal/metal oxide support interaction. For example [63], when IrO_2 nanoparticles are supported on antimony-doped tin oxide (ATO), the d band vacancy and iridium oxidation decreased due to the interfacial electron charge donation from ATO to iridium, evidenced by the shift of X-ray absorption white line toward lower energy. This is corroborated with the increased Ir-O bonding distances for ATO-supported IrOx from extended X-ray fine structure (EXAFS) analysis, which is consistent with a lower average iridium oxidation state. Such interaction between IrOx and ATO suppressed the growth of higher valent IrOx layer that leads to iridium dissolution, thereby improved catalyst stability [63]. Similar interaction between metal/metal oxide supports was also found for IrO_2 supported on TiO_2 catalyst where the catalyst stability was enhanced as well [64].

3.2. Mixed bimetallic oxide

Mixed bimetallic oxides of iridium and a non-noble metal have been used to optimize the anode catalyst for PEM electrolyzer application [65–73]. The most apparent benefit of bimetallic oxide is to reduce the iridium loading in the catalyst if the catalyst activity remains comparable or higher compared to pure IrOx. For example, at >10 mol% iridium content in IrO_2 + SnO_2 mixed oxide the Tafel slope for OER is identical with pure IrO_2, suggesting that the surface properties of IrO_2 + SnO_2 mixed oxide is dominated by IrO_2 and behave as pure IrO_2 [65]. In particular, iridium content in the 30–90 mol% range shows higher OER performance than pure IrO_2 [65].

More importantly, mixed bimetallic oxides modify the electronic or crystal structures of IrOx, which can significantly enhance the OER activity. The bond forming or breaking during OER is governed by the interaction between the O-2p orbital of intermediates with the d orbitals of surface sites of the transition metals [66–68, 70–73]. Thus, the OER activity depends on the d-orbital electronic structure of the transitional metals. For example, Sun et al. [73] doped IrOx with copper and obtained enhanced OER activity in acid media at 30–50 mol% concentration. The Cu doping led to an IrO_2 lattice distortion due to the CuO_6 octahedron's Jahn-Teller effect and also generated oxygen defects (**Figure 4a**), which significantly affected the energy distribution of the d-orbitals of Ir sites. The induced partial oxygen defects and the lifted degeneracy of t_{2g} and e_g orbitals reduced the energy required for the O-O bond formation, thereby enhanced the OER activity. In another case, Reier et al. [69] prepared Ir-Ni mixed oxide thin film and found a 20-fold enhancement of OER activity compared to pure IrOx thin film. As shown in **Figure 4b**, the surface Ni elements are leached out during OER and weakened the binding energy of the Ni-depleted oxygen with the lattice, forming oxygen with lower binding energy, and similar to the electrophilic oxygen. This promotes the formation of weakly bonded surface hydroxyls, which govern the overall OER reaction rate and suppress the formation of unreactive divalent = O species on the surface. Furthermore, Tae et al. [70] reported Ir-Co mixed oxide with 5% iridium

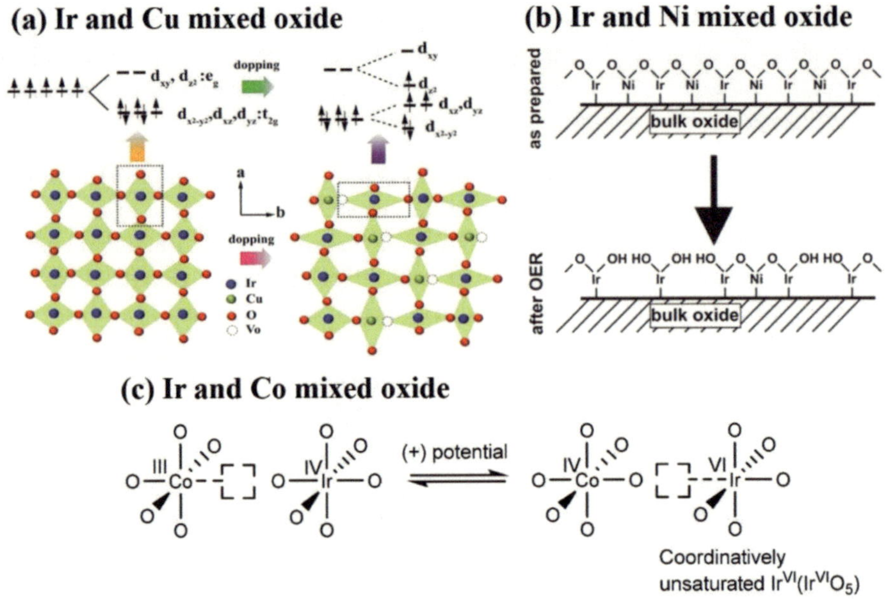

Figure 4. (a) Schematic lattice diagram in the ab plane of IrO_2 (left) and substituted by Cu (right). The top row shows Ir-5d orbitals degeneracy of IrO_2 (left) and the lift degeneracy and electron redistribution by doping with Cu [73]—Published by The Royal Society of Chemistry. (b) Model of Ni-leaching during OER for Ir-Ni mixed oxide surface, reproduced with permission [69]. Copyright (2015) American Chemical Society. (c) Transfer of oxygen vacancy from Co (III) to Ir (VI), reproduced with permission [70]. Copyright (2015) American Chemical Society.

loading that exhibits excellent OER activity and stability. At anodic potentials, the oxygen vacancy in Co (III) is transferred to Ir (IV) O_6 center, leading to the formation of coordinatively unsaturated Ir (VI) O_5 structure (**Figure 4c**), which is highly active for OER.

3.3. Effect of particle size

Small particles are usually favorable for higher mass activity due to the high surface area to bulk ratio, which facilitates the reduction of catalyst loading. More importantly, the change of electronic structure is unavoidably accompanied with the change of particle size, as evidenced in the early work on X-ray photoelectron spectroscopy (XPS) [74–76]. As particle size decreases, the binding energy is shifted to higher levels due to the increase of lattice strain and the coordination reduction [77, 78]. Richter et al. [78] shows that two mechanisms contribute to such an energy shift. The first is the initial state effect caused by the increase of lattice strain as particle size decreases, a result of d-hybridization; the second is the final state relaxation that increases with the decrease of particle size, which results in the stronger screening of the core hole leading to higher binding energy. Small particles show enhanced d-hybridization, which shift the d-band center to lower levels, decreasing the bond strength of the adsorbates during an electrochemical reaction [79].

Abbot et al. [34] systematically studied the particle size effect for IrOx in the range of 2 –30 nm. EXAFS analysis reveals that, as the IrOx particle size decreases, the coordination number decreases and the Ir-O bond length increases. The core-level binding energy was thus shifted to higher levels as the particle size decreases, which is in agreement with previous literature [74–76]. As a result, the iridium-oxygen bonding is weakened as the particle size decreases, leading to higher content of hydroxide, and lower average oxidation state of iridium from Ir (IV) to a combination of Ir (IV) and Ir (III). Based on previous discussions in Section 3.1, surface hydroxide and mixed Ir (IV) and Ir (III) state are favorable for OER. Indeed, Abbot et al. [34] show that the IrOx particles of 2 nm have 10 fold higher OER mass activity and three fold higher turnover frequency compared to the IrOx particles of 30 nm.

3.4. Effect of ionomer content

Perfluorosulfonic acid ionomer provides proton conductivity in the catalyst layer, extends the reaction zone, and improves the catalyst utilization [80, 81]. Ma et al. [82] focused on the effect of ionomer content on the ohmic resistance of PEM electrolyzers with iridium supported on titanium carbide (TiC) as an anode catalyst. Ma et al. [82] showed that increasing the ionomer content from 10 to 40 wt% causes a decrease in PEMWE performance. Using a simple non-linear fit:

$$E_{corrected} = E_{applied} - IR = a + b\ logI \tag{12}$$

The authors derived the ohmic loss contribution in the performance and concluded that higher ohmic resistance is associated with high ionomer content. It is noted that with this non-linear fit method, the obtained ohmic resistance includes contributions from all sources of the membrane electrode assembly (MEA) components, including the ionic resistance. Xu et al. [83] investigated the effect of ionomer content from the voltammetric charge and interfacial resistance perspective. The amount of voltammetric charges in cyclic voltammetry is proportional to the surface active sites [84]. A wider range of ionomer contents, from 5 to 40 wt%, were studied. The highest total charge was achieved at 25 wt% ionomer content, which is in agreement with the optimum PEMWE performance. Based on Butler-Volmer equation, the authors constructed a model to describe the I-V curves and concluded that the optimum ionomer content minimizes the interfacial resistance between the membrane and the electrode.

In a recent study, Bernt and Gasteiger [85] investigated the effect of anode ionomer content by analyzing the voltage loss contribution to the PEMWE performance. The ohmic resistance was found to increase abruptly when the ionomer content is above 20 wt%. This suggests that an electronically insulating film of residual ionomer forms at the electrode/GDL interface when the ionomer volume exceeds the void volume of the catalyst layer, which results in higher contact resistances. Further, such electronic insulation caused higher OER overpotential with >20 wt% ionomer content and thereby decreased the catalyst utilization. After subtracting the voltage losses due to ohmic and kinetic losses, Bernt and Gasteiger [85] attributed to the remaining losses (red diamond) to mass transport (**Figure 5**). The main path for oxygen removal from the electrode is not permeation but convective transport through the void volume of the anode. Thus, higher ionomer content imposes higher transport resistance for oxygen removal.

In sum, similar to PEM fuel cell electrodes, the ionomer content is an important parameter for the optimization of PEM electrolyzers. While low ionomer content provides insufficient proton

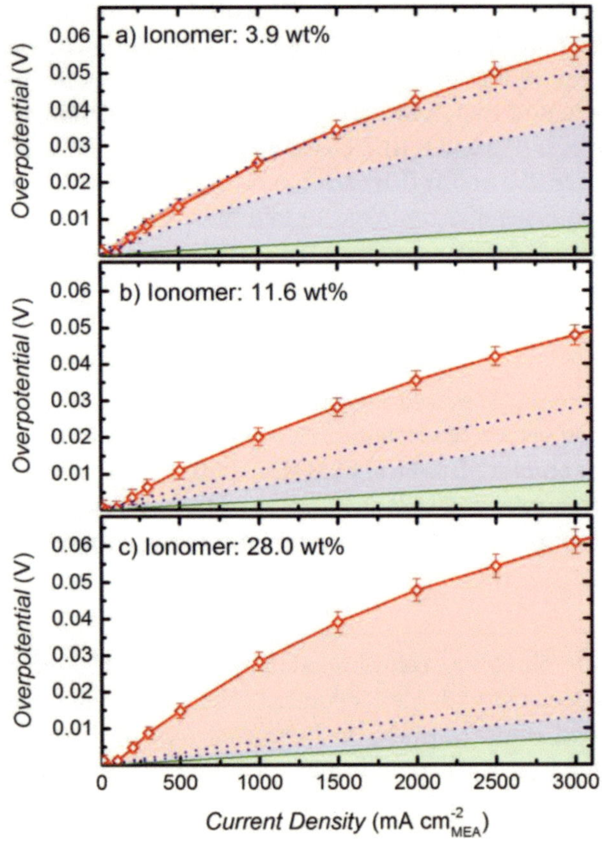

Figure 5. Overpotentials of PEM electrolyzer MEAs with different anode ionomer content after subtraction of ohmic and kinetics losses. Overpotentials from different sources are represented as different colors. Blue: anode proton conduction resistance; Green: cathode proton conduction resistance; Red: remaining overpotentials due to mass transport loss. Reproduced with permission [85]. Copyright (2016) The Electrochemical Society.

conduction and less accessible surface active sites, high ionomer content induces high interfacial resistance and mass transport loss due to insufficient void volume for oxygen diffusion. PEM electrolyzers are economically beneficial for a large-scale application operating at high current densities (>2 A cm^{-2}) and high pressures (30–45 bar). Thus, optimization of the ionomer content becomes crucial to minimize the cell potential loss at high current and pressure and improve cell efficiency.

4. PEM water electrolysis for energy storage

4.1. Integration with renewable energy technologies

Power generated from renewable energy sources, such as solar and wind, requires energy storage devices to balance its fluctuation and intermittence because of variable weather conditions [86]. Hydrogen production by water electrolysis has been developed as an alternative technology for energy conversion and storage that can be fitted to renewable energy systems [87, 88]. This section will briefly introduce the role of PEM electrolyzers in power-to-gas, solar, and wind energy systems.

Power-to-gas is emerging as a novel energy storage method that uses the surplus electricity from the grid during off-peak periods and converts it to hydrogen through a water electrolysis process [89]. The key technology for this strategy is the electrolyzer, which bridges the power to utilization by producing hydrogen. Electrolyzers must meet the following requirements in order for power-to-gas to become efficient and economically viable [90]: (1) high efficiency of hydrogen production; (2) fast response to power fluctuation; (3) very low minimal load for stand-by; (4) high-pressure operation to reduce the cost of hydrogen compression; and (5) long durability and lifetime. The major drawback of power-to-gas is the low efficiency and high cost of electrolyzers. While PEM electrolyzers offer fast response, high pressure, and the production of pure hydrogen, scaling up to MW scale has been the major technical challenge. As mentioned previously, the major hurdle of this challenge is the cost of catalysts and other MEA components (e.g., the Ti-based bipolar plates).

Solar energy conversion into hydrogen by water splitting has been long studied by various research groups due to its easy scale-up nature. The state-of-the-art photocatalytic and photoelectrochemical system has an efficiency of 10 and 5%, respectively, in the direct conversion of solar to hydrogen (STH) [91]. To reduce the price of hydrogen and boost up the efficiency of STH, an electrolyzer cell (EC) can be coupled with a photovoltaic cell (PV), called photovoltaic–water electrolysis system (PV-EC) [92]. The STH efficiency can be increased up to 30%. Further, a direct current to direct current (DC-DC) converter can be implemented between PV and EC (**Figure 6**). The main role of the converter is maximum power point tracking of the sun light to the EC system. Thus, coupling electrolyzers with highly efficient PVs and with

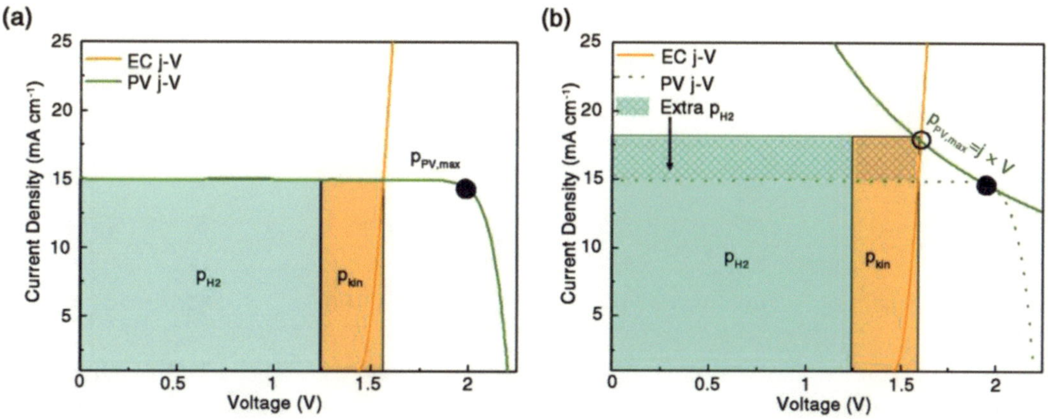

Figure 6. Design principle of the PV-Conv-EC system based on an independent PV, the EC performance, and the existence of a converter. (a) Hydrogen power per square centimeter (p_{H2}) and kinetic loss per square centimeter (p_{kin}) at a given current density–voltage (j − V) curve of the PV and EC. The intersection between the PV and EC j − V curve has a lower voltage and a higher current density than the $p_{PV,\ max\ point}$. (b) p_{H2} and p_{kin} after the DC-DC converter assistance on (a). Reproduced with permission [92]. Copyright (2015) American Chemical Society.

converter assistance, high STH efficiency PV-Conv-EC systems can achieve 20.6% STH efficiency and 78% PV electricity to hydrogen conversion efficiency [92].

Wind power is heavily influenced by meteorological variances and requires balancing power for the load fluctuation. Wind to hydrogen (WTH) strategies by water electrolysis could offer a solution to this problem [9]. Synergy between wind farms and water electrolyzers makes hydrogen a buffer solution to balance the grid power as well as produce hydrogen from surplus wind power [9]. Thus, the wind power can be utilized at its maximum capacity.

4.2. Current status of commercial manufacturers

Table 2 summarizes the commercially available PEM electrolysis systems from 12 major manufacturers worldwide. Note that only the highest capacity from each manufacturer is listed here. Six of the manufacturers are able to achieve a system capacity around 1 MW or higher. The largest system so far is from Hydrogenics, a 15 MW plant with 10 cell stacks capable of producing 3000 Nm^3 hydrogen per hour. Giner currently offers a more powerful cell stack, Kennebec stacks, even though its current systems have not yet incorporated this brand of the cell stack. The Kennebec stacks have the largest capacity among all the stacks available from these manufacturers. It has a capacity of 5 MW and is capable of producing 2200 kg of hydrogen per day (that is, 1020 Nm^3 per hour). Most of the systems offer high-purity hydrogen with the high delivery pressure that is suitable for on-site storage and fueling. The system consumption of electricity is very close among all the manufacturers, averaged around 5.5 ± 0.5 kWh per Nm^3 of hydrogen.

Manufacturers	System model	H2 production rate, Nm^3/hr	H2 purity (after purification)	Installed power, MW	System consumption, kWh/Nm^3 of H_2	Delivery pressure, bar	References
Hydrogenics	HyLYZER-3000	3000	99.998%	15	5.0–5.4	30	[93]
Proton Onsite	M400	400	>99.9995%	2	5	30	[94]
Siemens	Silyzer 200	225	99.9%	1.25	5.1–5.4	35	[95]
ITM power	HGas1000	215	99.999%	1.03	5.5	20–80	[96]
Giner	200S	200	99.999%	1	5	40	[97]
AREVA H2Gen	E120	120	99.999%	0.96	4.8	30	[98]
H-TEC	ME100/350	66	99.999%	0.225	4.9	20	[99]
Kobelco-eco solutions	SH60D	60	99.9999%	0.2	5.5–6.5	8.2	[100]
Treadwell Corp.	NA	10.2	NA	NA	NA	75.8	[101]
Angstrom Advanced	HGH170000	10	99.9999%	0.058	5.5	4	[102]
SylaTech	HE32	2	99.999%	0.01	4.9	30	[103]
GreenHydrogen	HyProvide P1	1	99.995%	0.01	5.5	50	[104]

Table 2. Summary of commercial PEM electrolysis systems.

5. Concluding remarks and outlook

The growing demand for energy and the accompanied environmental issues call for a rapid transition to low-carbon/carbon-free energy structure. In this context, hydrogen serves as an ideal secondary energy source for energy storage and transport. The key technology for hydrogen energy is water electrolysis. In particular, PEM electrolysis has been driven strongly by flexible energy storage in recent years as it offers several advantages compared to alkaline and solid oxide electrolysis. Nowadays, more mega-Watt scale PEM electrolysis systems are available on the market and in the field. However, further technological advancement is still demanded in the field of electrocatalysis and material science to obtain a deeper understanding of catalytic reactions and design new catalysts such that PEM electrolysis is more durable and cost-effective. Furthermore, as PEM electrolysis is but one building block for the future hydrogen economy, efforts in R&D should emphasize the compatibility with other technologies and optimize the synergic effects.

Conflict of interest

The authors have no conflict of interest to declare.

Author details

Radenka Maric[1,2] and Haoran Yu[1,2]*

*Address all correspondence to: haoran.yu@uconn.edu

1 Department of Chemical and Biomolecular Engineering, University of Connecticut, Storrs, CT, USA

2 Center for Clean Energy Engineering, University of Connecticut, Storrs, CT, USA

References

[1] Tverberg G. BP data suggests we are reaching peak energy demand. 2015. Available from: https://ourfiniteworld.com/2015/06/23/bp-data-suggests-we-are-reaching-peak-energy-demand [Accessed: April 14, 2018]

[2] Learning about renewable energy. Available from: https://www.nrel.gov/workingwithus/learning.html [Accessed: April 14, 2018]

[3] Vivas FJ, De las Heras A, Segura F, Andújar JM. A review of energy management strategies for renewable hybrid energy systems with hydrogen backup. Renewable and Sustainable Energy Reviews. 2018;**82**:126-155

[4] Winsche WE, Hoffman KC, Salzano FJ. Hydrogen: Its Future Role in the Nation's Energy Economy. Science. 1973;**180**:1325

[5] Bennaceur K, Clark B, Franklin MJ, Ramakrshnan TS, Roulet C, Stout E. Hydrogen: A future energy carrier? Oilfield Review. 2005;**17**:30-41

[6] Mazloomi K, Gomes C. Hydrogen as an energy carrier: Prospects and challenges. Renewable and Sustainable Energy Reviews. 2012;**16**:3024-3033

[7] Cipriani G, Di Dio V, Genduso F, La Cascia D, Liga R, Miceli R, et al. Perspective on hydrogen energy carrier and its automotive applications. International Journal of Hydrogen Energy. 2014;**39**:8482-8494

[8] Sharma S, Ghoshal SK. Hydrogen the future transportation fuel: From production to applications. Renewable and Sustainable Energy Reviews. 2015;**43**:1151-1158

[9] Hosseini SE, Wahid MA. Hydrogen production from renewable and sustainable energy resources: Promising green energy carrier for clean development. Renewable and Sustainable Energy Reviews. 2016;**57**:850-866

[10] Hanley ES, Deane J, Gallachóir BÓ. The role of hydrogen in low carbon energy futures—A review of existing perspectives. Renewable and Sustainable Energy Reviews. 2018;**82**: 3027-3045

[11] Buttler A, Spliethoff H. Current status of water electrolysis for energy storage, grid balancing and sector coupling via power-to-gas and power-to-liquids: A review. Renewable and Sustainable Energy Reviews. 2018;**82**:2440-2454

[12] Marshall A, Børresen B, Hagen H, Tsypkin M, Tunold R. Hydrogen production by advanced proton exchange membrane (PEM) water electrolysers—Reduced energy consumption by improved electrocatalysis. Energy. 2007;**32**:431-436

[13] Millet P, Ngameni R, Grigoriev SA, Mbemba N, Brisset F, Ranjbari A, et al. PEM water electrolyzers: From electrocatalysis to stack development. International Journal of Hydrogen Energy. 2010;**35**:5043-5052

[14] Millet P, Mbemba M, Grigoriev SA, Fateev VN, Aukauloo A, Etievant C. Electrochemical performances of PEM water electrolysis cells and perspectives. International Journal of Hydrogen Energy. 2011;**36**:4134-4142

[15] Carmo M, Fritz DL, Mergel J, Stolten D. A comprehensive review on PEM water electrolysis. International Journal of Hydrogen Energy. 2013;**38**:4901-4934

[16] Zafar S. Overview of Biomass Pyrolysis. 2018. Available from: https://www.bioenergyconsult.com/biomass-pyrolysis/ [Accessed: April 15, 2018]

[17] Møller KT, Jensen TR, Akiba E, Li H. Hydrogen - A sustainable energy carrier. Progress in Natural Science: Materials International. 2017;**27**:34-40

[18] Lototskyy MV, Tolj I, Pickering L, Sita C, Barbir F, Yartys V. The use of metal hydrides in fuel cell applications. Progress in Natural Science: Materials International. 2017;**27**:3-20

[19] Chen P, Zhu M. Recent progress in hydrogen storage. Materials Today. 2008;**11**:36-43

[20] Ursua A, Gandia LM, Sanchis P. Hydrogen production from water electrolysis: Current status and future trends. Proceedings of the IEEE. 2012;**100**:410-426

[21] Zhang Y, Lundblad A, Campana PE, Yan J. Comparative study of battery storage and hydrogen storage to increase photovoltaic self-sufficiency in a residential building of Sweden. Energy Procedia. 2016;**103**:268-273

[22] Fabbri E, Habereder A, Waltar K, Kotz R, Schmidt TJ. Developments and perspectives of oxide-based catalysts for the oxygen evolution reaction. Catalysis Science & Technology. 2014;**4**:3800-3821

[23] Reier T, Nong HN, Teschner D, Schlogl R, Strasser P. Electrocatalytic oxygen evolution reaction in acidic environments: Reaction mechanisms and catalysts. Advanced Energy Materials. 2017;**7**. 1601275

[24] Ayers KE, Anderson EB, Capuano C, Carter B, Dalton L, Hanlon G, et al. Research advances towards low cost, high efficiency PEM electrolysis. ECS Transactions. 2010;**33**:3-15

[25] Danilovic N, Ayers KE, Capuano C, Renner JN, Wiles L, Pertoso M. (Plenary) Challenges in going from laboratory to megawatt scale PEM electrolysis. ECS Transactions. 2016;**75**:395-402

[26] Spori C, Kwan JTH, Bonakdarpour A, Wilkinson DP, Strasser P. The stability challenges of oxygen evolving catalysts: Towards a common fundamental understanding and mitigation of catalyst degradation. Angewandte Chemie International Edition. 2017;**56**:5994-6021

[27] Danilovic N, Subbaraman R, Chang KC, Chang SH, Kang Y, Snyder J, et al. Using surface segregation to design stable Ru-Ir oxides for the oxygen evolution reaction in acidic environments. Angewandte Chemie International Edition. 2014;**53**:14016-14021

[28] Danilovic N, Subbaraman R, Chang K, Chang SH, Kang YJ, Snyder J, et al. Activity-stability trends for the oxygen evolution reaction on monometallic oxides in acidic environments. Journal of Physical Chemistry Letters. 2014;**5**:2474-2478

[29] Fierro S, Kapałka A, Comninellis C. Electrochemical comparison between IrO_2 prepared by thermal treatment of iridium metal and IrO_2 prepared by thermal decomposition of H_2IrCl_6 solution. Electrochemistry Communications. 2010;**12**:172-174

[30] Reier T, Teschner D, Lunkenbein T, Bergmann A, Selve S, Kraehnert R, et al. Electrocatalytic oxygen evolution on iridium oxide: Uncovering catalyst-substrate interactions and active iridium oxide species. Journal of the Electrochemical Society. 2014;**161**:F876-F882

[31] Geiger S, Kasian O, Shrestha BR, Mingers AM, Mayrhofer KJJ, Cherevko S. Activity and stability of electrochemically and thermally treated iridium for the oxygen evolution reaction. Journal of the Electrochemical Society. 2016;**163**:F3132-F3138

[32] Cherevko S, Reier T, Zeradjanin AR, Pawolek Z, Strasser P, Mayrhofer KJJ. Stability of nanostructured iridium oxide electrocatalysts during oxygen evolution reaction in acidic environment. Electrochemistry Communications. 2014;**48**:81-85

[33] Xu J, Wang M, Liu G, Li J, Wang X. The physical–chemical properties and electrocatalytic performance of iridium oxide in oxygen evolution. Electrochimica Acta. 2011;**56**:10223-10230

[34] Abbott DF, Lebedev D, Waltar K, Povia M, Nachtegaal M, Fabbri E, et al. Iridium oxide for the oxygen evolution reaction: Correlation between particle size, morphology, and the surface hydroxo layer from operando XAS. Chemistry of Materials. 2016;**28**:6591-6604

[35] Chandra D, Takama D, Masaki T, Sato T, Abe N, Togashi T, et al. Highly efficient electrocatalysis and mechanistic investigation of intermediate IrOx(OH)y nanoparticle films for water oxidation. ACS Catalysis. 2016;**6**:3946-3954

[36] Pfeifer V, Jones TE, Velasco Velez JJ, Massue C, Arrigo R, Teschner D, et al. The electronic structure of iridium and its oxides. Surface and Interface Analysis. 2016;**48**:261-273

[37] Minguzzi A, Lugaresi O, Achilli E, Locatelli C, Vertova A, Ghigna P, et al. Observing the oxidation state turnover in heterogeneous iridium-based water oxidation catalysts. Chemical Science. 2014;**5**:3591-3597

[38] Minguzzi A, Locatelli C, Lugaresi O, Achilli E, Cappelletti G, Scavini M, et al. Easy Accommodation of different oxidation states in iridium oxide nanoparticles with different hydration degree as water oxidation electrocatalysts. ACS Catalysis. 2015;**5**:5104-5115

[39] Cherevko S, Geiger S, Kasian O, Mingers A, Mayrhofer KJJ. Oxygen evolution activity and stability of iridium in acidic media. Part 2. – Electrochemically grown hydrous iridium oxide. Journal of Electroanalytical Chemistry. 2016;**774**:102-110

[40] Grigoriev SA, Bessarabov DG, Fateev VN. Degradation mechanisms of MEA characteristics during water electrolysis in solid polymer electrolyte cells. Russian Journal of Electrochemistry. 2017;**53**:318-323

[41] Rozain C, Mayousse E, Guillet N, Millet P. Influence of iridium oxide loadings on the performance of PEM water electrolysis cells: Part II. – Advanced oxygen electrodes, Applied Catalysis B: Environmental. 2016;**182**:123-131

[42] Siracusano S, Baglio V, Grigoriev SA, Merlo L, Fateev VN, Aricò AS. The influence of iridium chemical oxidation state on the performance and durability of oxygen evolution catalysts in PEM electrolysis. Journal of Power Sources. 2017;**366**:105-114

[43] Rakousky C, Reimer U, Wippermann K, Carmo M, Lueke W, Stolten D. An analysis of degradation phenomena in polymer electrolyte membrane water electrolysis. Journal of Power Sources. 2016;**326**:120-128

[44] Lettenmeier P, Wang R, Abouatallah R, Helmly S, Morawietz T, Hiesgen R, et al. Durable membrane electrode assemblies for proton exchange membrane electrolyzer systems operating at high current densities. Electrochimica Acta. 2016;**210**:502-511

[45] Lewinski KA, van dV, Luopa SM. NSTF advances for PEM electrolysis - the effect of alloying on activity of NSTF electrolyzer catalysts and performance of NSTF based PEM electrolyzers. ECS Transactions. 2015;**69**:893-917

[46] Wang L, Saveleva VA, Zafeiratos S, Savinova ER, Lettenmeier P, Gazdzicki P, et al. Highly active anode electrocatalysts derived from electrochemical leaching of Ru from metallic $Ir_{0.7}Ru_{0.3}$ for proton exchange membrane electrolyzers. Nano Energy. 2017;**34**: 385-391

[47] Siracusano S, Hodnik N, Jovanovic P, Ruiz-Zepeda F, Šala M, Baglio V, et al. New insights into the stability of a high performance nanostructured catalyst for sustainable water electrolysis. Nano Energy. 2017;**40**:618-632

[48] Liu G, Xu J, Wang Y, Wang X. An oxygen evolution catalyst on an antimony doped tin oxide nanowire structured support for proton exchange membrane liquid water electrolysis. Journal of Materials Chemistry A. 2015;**3**:20791-20800

[49] Damjanovic A, Dey A, Bockris JO. Kinetics of oxygen evolution and dissolution on platinum electrodes. Electrochimica Acta. 1966;**11**:791-814

[50] Damjanovic A, Wong MKY. On the mechanism of oxygen evolution at iridium electrodes. Journal of The Electrochemical Society. 1967;**114**:592-593

[51] Busch M, Ahlberg E, Panas I. Electrocatalytic oxygen evolution from water on a Mn(iii-v) dimer model catalyst-A DFT perspective. Physical Chemistry Chemical Physics. 2011;**13**: 15069-15076

[52] Steegstra P, Busch M, Panas I, Ahlberg E. Revisiting the redox properties of hydrous iridium oxide films in the context of oxygen evolution. Journal of Physical Chemistry C. 2013;**117**:20975-20981

[53] Cherevko S, Geiger S, Kasian O, Mingers A, Mayrhofer KJJ. Oxygen evolution activity and stability of iridium in acidic media. Part 1. – Metallic iridium. Journal of Electroanalytical Chemistry. 2016;**773**:69-78

[54] Xu D, Diao P, Jin T, Wu Q, Liu X, Guo X, et al. Iridium oxide nanoparticles and iridium/ iridium oxide nanocomposites: Photochemical fabrication and application in catalytic reduction of 4-nitrophenol. ACS Applied Materials & Interfaces. 2015;**7**:16738-16749

[55] Cruz AM, Abad L, Carretero NM, Moral-Vico J, Fraxedas J, Lozano P, et al. Iridium oxohydroxide, a significant member in the family of iridium oxides. Stoichiometry, characterization, and implications in bioelectrodes. Journal of Physical Chemistry C. 2012;**116**:5155-5168

[56] Sanchez Casalongue HG, Ng ML, Sarp K, Daniel F, Hirohito O, Anders N. In Situ observation of surface species on iridium oxide nanoparticles during the oxygen evolution reaction. Angewandte Chemie International Edition. 2014;**53**:7169-7172

[57] Pfeifer V, Jones TE, Velasco Velez JJ, Arrigo R, Piccinin S, Havecker M, et al. In situ observation of reactive oxygen species forming on oxygen-evolving iridium surfaces. Chemical Science. 2017;**8**:2143-2149

[58] Massue C, Pfeifer V, van Gastel M, Noack J, Algara-Siller G, Cap S, et al. Reactive electrophilic O^{I-} species evidenced in high-performance iridium oxohydroxide water oxidation electrocatalysts. ChemSusChem. 2017;**10**:4786-4798

[59] Pfeifer V, Jones TE, Wrabetz S, Massue C, Velasco Velez JJ, Arrigo R, et al. Reactive oxygen species in iridium-based OER catalysts. Chemical Science. 2016;**7**:6791-6795

[60] Cyriac M, Verena P, Xing H, Johannes N, Andrey T, Sébastien C, et al. High-performance supported iridium oxohydroxide water oxidation electrocatalysts. ChemSusChem. 2017; **10**:1943-1957

[61] Pfeifer V, Jones TE, Velasco Velez JJ, Massue C, Greiner MT, Arrigo R, et al. The electronic structure of iridium oxide electrodes active in water splitting. Physical Chemistry Chemical Physics. 2016;**18**:2292-2296

[62] Bernicke M, Ortel E, Reier T, Bergmann A, Ferreira de Araujo J, Strasser P, et al. Iridium oxide coatings with templated porosity as highly active oxygen evolution catalysts: Structure-activity relationships. ChemSusChem. 2015;**8**:1908-1915

[63] Oh HS, Nong HN, Reier T, Bergmann A, Gliech M, Ferreira de Araújo J, et al. Electrochemical catalyst – support effects and their stabilizing role for IrOx nanoparticle catalysts during the oxygen evolution reaction. Journal of the American Chemical Society. 2016;**138**:12552-12563

[64] Oakton E, Lebedev D, Povia M, Abbott DF, Fabbri E, Fedorov A, et al. IrO_2-TiO_2: A high-surface-area, active, and stable electrocatalyst for the oxygen evolution reaction. ACS Catalysis. 2017;**7**:2346-2352

[65] De Pauli CP, Trasatti S. Composite materials for electrocatalysis of O_2 evolution: IrO_2+SnO_2 in acid solution. Journal of Electroanalytical Chemistry. 2002;**538–539**:145-151

[66] Petrykin V, Macounová K, Okube M, Mukerjee S, Krtil P. Local structure of Co doped RuO_2 nanocrystalline electrocatalytic materials for chlorine and oxygen evolution. Catalysis Today. 2013;**202**:63-69

[67] Kadakia K, Datta MK, Velikokhatnyi OI, Jampani PH, Kumta PN. Fluorine doped (Ir,Sn, Nb)O_2 anode electro-catalyst for oxygen evolution via PEM based water electrolysis. International Journal of Hydrogen Energy. 2014;**39**:664-674

[68] González-Huerta RG, Ramos-Sánchez G, Balbuena PB. Oxygen evolution in Co-doped RuO_2 and IrO_2: Experimental and theoretical insights to diminish electrolysis overpotential. Journal of Power Sources. 2014;**268**:69-76

[69] Reier T, Pawolek Z, Cherevko S, Bruns M, Jones T, Teschner D, et al. Molecular insight in structure and activity of highly efficient, low-Ir Ir-Ni oxide catalysts for electrochemical water splitting (OER). Journal of the American Chemical Society. 2015;**137**:13031-13040

[70] Tae EL, Song J, Lee AR, Kim CH, Yoon S, Hwang IC, et al. Cobalt oxide electrode doped with iridium oxide as highly efficient water oxidation electrode. ACS Catalysis. 2015;**5**: 5525-5529

[71] Tran V, Yatabe T, Matsumoto T, Nakai H, Suzuki K, Enomoto T, et al. An IrSi oxide film as a highly active water-oxidation catalyst in acidic media. Chemical Communications. 2015;**51**: 12589-12592

[72] Yu A, Lee C, Kim MH, Lee Y. Nanotubular iridium-cobalt mixed oxide crystalline architectures inherited from cobalt oxide for highly efficient oxygen evolution reaction catalysis. ACS Applied Materials & Interfaces. 2017;**9**:35057-35066

[73] Sun W, Song Y, Gong X, Cao L, Yang J. An efficiently tuned d-orbital occupation of IrO_2 by doping with Cu for enhancing the oxygen evolution reaction activity. Chemical Science. 2015;**6**:4993-4999

[74] Mason MG, Gerenser LJ, Lee S-T. Electronic structure of catalytic metal clusters studied by x-ray photoemission spectroscopy. Physical Review Letters. 1977;**39**:288-291

[75] Mason MG. Electronic structure of supported small metal clusters. Physical Review B. 1983;**27**:748-762

[76] Wertheim GK, DiCenzo SB, Buchanan DNE. Noble- and transition-metal clusters: The d bands of silver and palladium. Physical Review B. 1986;**33**:5384-5390

[77] Sun CQ. Surface and nanosolid core-level shift: Impact of atomic coordination-number imperfection. Physical Review B. 2004;**69**:045105

[78] Richter B, Kuhlenbeck H, Freund H-J, Bagus PS. Cluster core-level binding-energy shifts: The role of lattice strain. Physical Review Letters. 2004;**93**:026805

[79] Zhou WP, Lewera A, Larsen R, Masel RI, Bagus PS, Wieckowski A. Size effects in electronic and catalytic properties of unsupported palladium nanoparticles in electrooxidation of formic acid. The Journal of Physical Chemistry. B. 2006;**110**:13393-13398

[80] Litster S, McLean G. PEM fuel cell electrodes. Journal of Power Sources. 2004;**130**:61-76

[81] Holdcroft S. Fuel cell catalyst layers: A polymer science perspective. Chemistry of Materials. 2014;**26**:381-393

[82] Ma L, Sui S, Zhai Y. Investigations on high performance proton exchange membrane water electrolyzer. International Journal of Hydrogen Energy. 2009;**34**:678-684

[83] Xu W, Scott K. The effects of ionomer content on PEM water electrolyser membrane electrode assembly performance. International Journal of Hydrogen Energy. 2010;**35**: 12029-12037

[84] Ma H, Liu C, Liao J, Su Y, Xue X, Xing W. Study of ruthenium oxide catalyst for electrocatalytic performance in oxygen evolution. Journal of Molecular Catalysis A: Chemical. 2006;**247**:7-13

[85] Bernt M, Gasteiger HA. Influence of ionomer content in IrO_2/TiO_2 electrodes on PEM water electrolyzer performance. Journal of the Electrochemical Society. 2016;**163**:F3179-F3189

[86] Park S, Shao Y, Liu J, Wang Y. Oxygen electrocatalysts for water electrolyzers and reversible fuel cells: status and perspective. Energy & Environmental Science. 2012;**5**: 9331-9344

[87] Holladay JD, Hu J, King DL, Wang Y. An overview of hydrogen production technologies. Catalysis Today. 2009;**139**:244-260

[88] Arico AS, Siracusano S, Briguglio N, Baglio V, Di Blasi A, Antonucci V. Polymer electrolyte membrane water electrolysis: status of technologies and potential applications in combination with renewable power sources. Journal of Applied Electrochemistry. 2013; **43**:107-118

[89] Götz M, Lefebvre J, Mörs F, McDaniel Koch A, Graf F, Bajohr S, et al. Renewable power-to-gas: A technological and economic review. Renewable Energy. 2016;**85**:1371-1390

[90] Schiebahn S, Grube T, Robinius M, Tietze V, Kumar B, Stolten D. Power to gas: Technological overview, systems analysis and economic assessment for a case study in Germany. International Journal of Hydrogen Energy. 2015;**40**:4285-4294

[91] Chang WJ, Lee KH, Ha JI, Nam KT. Hydrogen production via water electrolysis: The benefits of a solar cell-powered process. IEEE Electrification Magazine. 2018;**6**:19-25

[92] Chang WJ, Lee K, Ha H, Jin K, Kim G, Hwang S, et al. Design principle and loss engineering for photovoltaic-electrolysis cell system. ACS Omega. 2017;**2**:1009-1018

[93] Hydrogenics. Hydrogenics Renewable Hydrogen Solutions. Available from: http://www.hydrogenics.com/wp-content/uploads/Renewable-Hydrogen-Brochure.pdf [Accessed: April 12, 2018]

[94] Proton Onsite. High Capacity Hydrogen Systems: M Series PEM Electrolyzers. Available from: http://www.protononsite.com/sites/default/files/2016-10/pd-0600-0114_rev_a.pdf [Accessed: April 14, 2018]

[95] Siemens. Available from: https://www.industry.siemens.com/topics/global/en/pem-electrolyzer/silyzer/silyzer-systems/Pages/silyzer-systeme1105-9496.aspx [Accessed: April 14, 2018]

[96] ITM-power. Available from: http://www.itm-power.com/product/hgas [Accessed: April 14, 2018]

[97] Giner. Available from: https://www.ginerelx.com/electrolyzer-systems [Accessed: April 14, 2018]

[98] Arevah2gen. Available from: http://www.arevah2gen.com/wpcontent/uploads/2017/02/AREVA_H2GEN_FICHE_PRODUIT_FRANCE_HD.pdf [Accessed: April 14, 2018]

[99] H-TEC. Available from: https://www.h-tec-systems.com/fileadmin/Content/PDFs/24022018/HTEC_SYSTEMS_Datasheet_Series-ME_24022018.pdf [Accessed: April 14, 2018]

[100] Kobelco-eco. Available from: http://www.kobelco-eco.co.jp/product/suisohassei/img/catalog_hhog.pdf. [Accessed: April 14, 2018]

[101] Treadwell Corp. Available from: http://www.treadwellcorp.com/proton-exchange-membrane-hydrogen-generator.php [Accessed: April 14, 2018]

[102] Angstrom-advanced. Available from: http://www.angstrom-advanced.com/index.asp?page=HGH10000 [Accessed: April 14, 2018]

[103] Sylatech. Available from: http://www.sylatech.de/pdf/Elektrolyseure.pdf [Accessed: April 14, 2018]

[104] Greenhydrogen. Available from: http://greenhydrogen.dk/technology/hyprovide/ [Accessed: April 14, 2018]

Chapter 7

Nanostructured Si/SiO_2 Quantum Wells

Toshio Takeuchi and Yoshiji Horikoshi

Additional information is available at the end of the chapter

http://dx.doi.org/10.5772/intechopen.79880

Abstract

The motivation for developing light-emitting devices on an indirect transition semiconductor such as silicon has been widely discussed for Si/SiO_2 nanostructures. In this chapter, we report on the fabrication of Si/SiO_2 quantum-confined amorphous nanostructured films and their optical properties. The Si/SiO_2 nanostructures comprising amorphous Si, SiO_2, and Si/SiO_2 multilayers are grown using ultrahigh vacuum radio frequency magnetron sputtering. Optical absorption coefficients of the Si/SiO_2 nanostructures are evaluated with regard to tentative integrated Si thicknesses. Optical energy band gaps of the Si/SiO_2 multilayer films are in accordance with the effective mass theory and described as $E_0 = 1.61 + 0.75d^{-2}$ eV at the Si layer-integrated thicknesses ranging from 0.5 to 6 nm. Quantum confinement effects in the Si/SiO_2 nanostructures are inferred from optical transmittance and reflectance spectra. The rapid-thermal-annealed Si/SiO_2 multilayer films demonstrate the intensified photoluminescence at ~1.45 eV due to the formation of nanocrystalline silicon. The temperature dependence of the nanocrystalline luminescence intensity shows the nonmonotonous behavior which is interpreted invoking the Kapoor model.

Keywords: amorphous Si/SiO_2, quantum confinement, nanocrystals, optical properties, absorption coefficient, photoluminescence, the Kapoor model

1. Introduction

Silicon, the principal semiconducting material, inherits the indirect optical transitions from its band structure. The research efforts are put forth on realization of light emission effects in silicon-based Si/SiO_2 nanostructured devices exploring hydrogenated amorphous Si [1–9], porous Si [10–12], Si quantum dots [13–19], amorphous Si quantum wells (QWs) [20–24], crystalline and nanocrystalline Si QWs [25–28], and Er-doped QWs [29–31]. The fabrication of Si/SiO_2 QWs has been an attractive area in process technology of Si-based light-emitting devices in last few decades. Various techniques are developed to synthesize the Si/SiO_2 nanostructured films—molecular beam epitaxy (MBE) [32–37], plasma-enhanced chemical vapor deposition (PECVD) [38–49], magnetron sputtering [50–65], electrochemical dissolution in electrolytes, ion implantation, and others. In this chapter, we investigate the growth of amorphous Si/SiO_2 QWs employing an ultrahigh vacuum (UHV) radio frequency (RF) magnetron sputtering (MS) system. This method is simple and easy to use in a manual operation. The Si/SiO_2 QW films are fabricated on sapphire and silicon wafers at room temperature to enable the atomic precision of the film growth via minimization of atomic movements during and after the depositions. Morphology, crystallinity, atomic bonding, and structures of the Si/SiO_2 films are evaluated by means of focused-ion-beam scanning electron microscopy (FIB-SEM), X-ray diffraction (XRD), and high-resolution X-ray photoelectron spectroscopy (XPS). The Si/SiO_2 films are distinguished with the layer number (the period), the Si thickness (the QW thickness), and the SiO_2 thickness (the barrier thickness). The first identification of the quantum confinement effects is made speculating on the optical energy band gap determined from the optical absorption and reflectance measurements taking into account the energy band gaps of silicon and fused quartz as ~1.1 and ~7.8 eV, respectively. To minimize the experimental uncertainty, the Si/SiO_2 films are deposited at room temperature. At the raised temperature conditions, the uncertainty remains the atomic diffusion, reactions, and oxidation at the Si/SiO_2 interface. On the other hand, the deposited Si and SiO_2 layers are expected as in amorphous conditions. At first, we restricted the Si/SiO_2 nanostructured layer films to be amorphous. Photoluminescence (PL) is used to characterize the as-grown and annealed materials. The thermal annealing is expected to improve the photoluminescence characteristics.

2. Experiment

2.1. Si/SiO_2 layer films preparation

Si/SiO_2 QWs films are synthesized in an ultrahigh vacuum (UHV; 3×10^{-8} Pa) RF MS system at a very small deposition rate (from 0.005 to 0.5 nm/s). The schematics of the UHV RF magnetron sputtering systems are shown in **Figure 1**. The ultrahigh vacuum chamber is equipped with two AJA A300 UHV RF magnetron sputtering guns connected to argon and oxygen gas lines, sputter ion guns, and 5 N Si and 5 N fused quartz SiO_2 targets. Preparation temperatures are controlled at the substrate holder. Transparent substrates are used for optical measurements; crystalline and amorphous substrates are used to test the influence of substrate crystallinity on the film growth. All depositions are operated at room temperature on both transparent sapphire A and opaque Si (100) substrates. The polished sapphire substrates are etched in dilute HF and put into the UHV chamber. The base pressure of the chamber is 10^{-7} Pa, and the sputtering gun pressure during the plasma operation in argon is 2×10^{-1} Pa. The deposition process is a repetition of Si and SiO_2 depositions separated by an interval time. The deposition parameters of the quantum-well structures are the well layer thickness, the barrier layer thickness, and the number of periodicity. The thickness of each layer is controlled by the deposition speed and the sputtering time. Basic deposition speeds are 0.05 nm/s for Si and

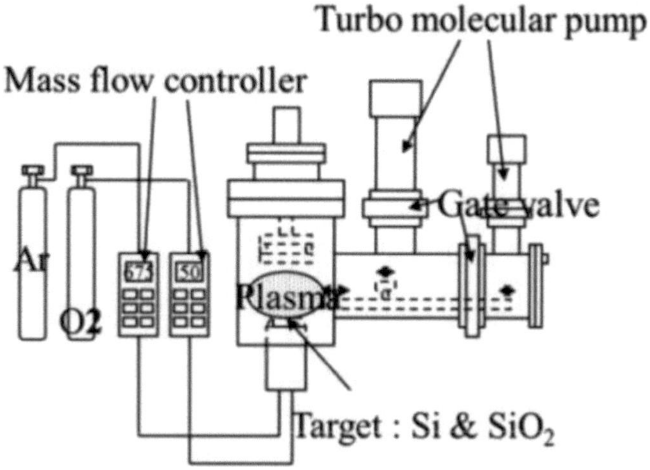

Figure 1. A schematic drawing of the ultrahigh vacuum radio frequency magnetron sputtering. Sputtering targets can be changed at each deposition program.

0.021 nm/s for SiO_2 at an argon pressure of 0.2 Pa. Minimization of the atomic diffusion and the oxidation during the depositions are the main concerns. The Si/SiO_2 nanostructure films are made of 1–50 periods consisting of 0.5–15 nm thick Si QWs and 0.5–6 nm thick SiO_2 barriers. The 10-layer Si/SiO_2 nanostructured layers are formed with the total thickness of 10–200 Å. **Figure 2** is the SEM cross-sectional view of a 20-layer Si/SiO_2 nanostructure comprising 2.0 nm Si and 2.1 nm SiO_2. It shows 20 pairs of the white thin SiO_2 layers and dark Si layers.

Figure 2. A cross-sectional view of an SEM photograph for an as-deposited 20-layer amorphous Si/SiO_2 film deposited on a fused quartz substrate.

2.1.1. Crystallinity

The XRD spectra (CuKα source) of 180-nm thick Si single layer and 150-nm thick SiO_2 single layer prepared on the sapphire substrates at room temperature do not show the crystallinity of the samples. Both spectra reveal the noncrystalline characteristics of Si and SiO_2 films.

2.1.2. Density-of-states structures

Atomic constitutions of each layer are evaluated with XPS on binding energy of Si2p and O1s electrons. The bulk Si2p core-level binding energy for Si(111) is ~99.3 eV and the bulk Si2p oxide binding energy value for SiO_2 is ~103.7 eV referring to Keister [66]. PHI 500 Versa Probe II scanning XPS microprobe is designed to take out a 10-degree signal, enable slow speed (0.01 nm/s SiO_2) area etching, and equipped with a monochromatic AlKα X-ray source. The depth profiles are characterized by using a low-energy argon ion gun to avoid selective etching. The binding energy dependence of the core densities of states at each etched depth suggests periodic distributions of each atomic composition. This analytical technique has particular applicability to the evaluation of the density of states with atomic contributions. **Figure 3(a)** is the plot of the density of states from 97 to 107 eV as the parameters of depth. The profiles of two peaks (a) 98.9 eV Si 2p spectra (Si^0) (element: un-oxidized) and the 103.2 eV Si 2p (oxidized) (Si^{4+}) are shown. In **Figure 3(b)**, 532.6 eV O1s single peaks are shown. The spectra (a) 103.2 eV and (b) 532.5 eV exhibit maxima at the same depth and spectrum (a) has a minimum at the bottom of (b). **Figure 3(a)**, at the depth of 0.54 nm, 103.2 eV intensity peaks and 98.9 eV show a minimum. The density-of-states depth profiles explain the presence of the Si/SiO_2-layered amorphous nanostructure fabricated using the UHV RF magnetron sputtering method at the atomic scale precision.

2.2. Optical properties of Si/SiO_2 layer films

Optical transmittance spectra and reflectance spectra are measured with the help of JASCO V-670 visible and ultraviolet optical photometer at room temperature. Optical properties of an

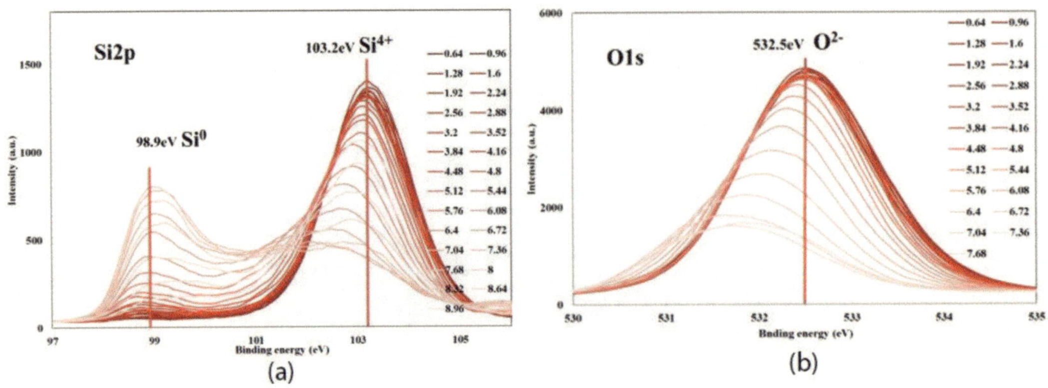

Figure 3. The density-of-states intensity of nanostructure Si/SiO_2 amorphous films. (a) Si2p 99.1 eV (elemental Si) and Si 2p 103.3 eV (oxidized Si) and (b) O1s 532.6 eV.

amorphous Si/SiO$_2$ nanostructure film show the higher optical transmittance and wide optical window effects. Unique optical properties are a candidate for solar windows in solar cells or filters of ultraviolet light. The parameters characterizing the Si/SO$_2$ film structures are the well layer thickness, the barrier layer thickness, and the number of periodicity. **Figure 4** displays the optical reflectance and transmittance spectra of amorphous Si/SiO$_2$-nanostructured layer films of various period numbers. The well thickness of the samples changes from 2 to 24 nm, while the barrier thickness is fixed at 4.8 nm. As the period's number of layers increases, the optical reflection decreases and the optical transmittance increases markedly, although the onset energy of transmittance and the absorption edge wavelength show the constant values. The increasing period number enhances optical transmittance and decreases optical reflectance. The spectra are saturating at 8–12 barrier layers. Increasing the period number does not change the absorption edge energy. Nanostructure effects observed on the 12-layer Si/SiO$_2$ film as the optical transmittance and reflectance effects are saturating. **Figure 5** exhibits the optical reflectance (a) and transmittance spectra (b) of Si/SiO$_2$ films as a function of the Si well thickness at the constant 12-period numbers and the constant barrier thickness of 4.8 nm. The increasing Si well thickness increases the reflectance and decreases the transmittance as expected. Also, the absorption edge energy shows the constant values. **Figure 6** shows the barrier thicknesses dependence of optical reflectance (a) and optical transmittance (b) spectra for the constant 12-layer period and 2 nm well thicknesses. The increasing barrier thicknesses diminish the optical reflectance and enhance the optical transmittance.

2.2.1. *Absorption coefficient*

Absorption coefficients α (λ) are used as the index of intrinsic properties of thin film materials. Absorption coefficient α (λ) at a wave length λ is evaluated from Eq. (1) for the sample thickness d with the optical transmittance T(λ) and reflectance R(λ). On the Si/SiO$_2$ nanostructure multilayer films, the integrated thickness of the Si layer, the reduced film thickness in Eq. (2) is used as the tentative thicknesses.

$$T = \frac{(1-R)^2 \exp(-\alpha d)}{1 - R^2 \exp(-2\alpha d)} \tag{1}$$

$$d = \sum_{i=0}^{n} d_i \tag{2}$$

$$(\alpha h\nu)^{1/2} = \beta(h\nu - E_0) \quad \alpha > 10^3 \tag{3}$$

Figure 7 shows the dependence of the absorption coefficients on the Si well thickness of 12-layer Si/SiO$_2$ films. The photon energy dependences of absorption coefficient show a sharp rise in the energy of absorption edges above 1000/cm. The dependence of the absorption edge energies on the Si well layer thickness is measured from 0.5 to 6 nm at the SiO$_2$ barrier layer fixed at 2.4 nm. In **Figure 8**, $(\alpha h\nu)^{1/2}$ vs. photon energy is plotted. The absorption coefficients of amorphous films are related in Eq. (3) known as an amorphous plot to obtain the absorption edge energy.

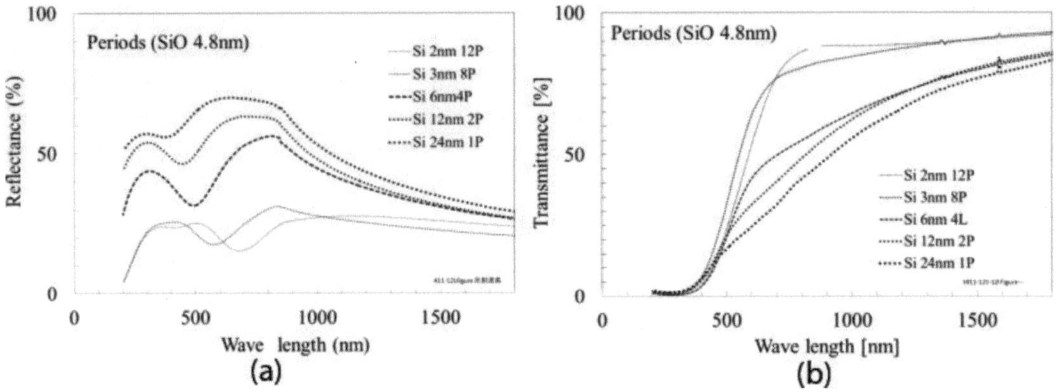

Figure 4. Optical properties of amorphous Si/SiO_2 multilayer films at the constant barrier thickness and the constant integrated well thicknesses as a function of the layer number. (a) Optical reflectance spectra, (b) optical transmittance spectra.

2.2.2. Quantum confinement

In **Figure 9**, the absorption edge energy is plotted vs. the tentative well thicknesses in a Si/SiO_2 multilayer structure and compared with the effective-mass theoretical estimations. Two types of the absorption edge energy evaluated from **Figures 7** and **8** are indicated. The absorption edge energy becomes larger as the QW thickness gets smaller. The blue shifts of the absorption energy are impressive in **Figure 9**. The absorption edge energy values E_0 are evaluated for each well thickness following the effective-mass theory, Eq. (4) [9]. The Si layer thickness dependency of absorption edge energy is in accordance with the effective-mass theory for thicknesses $0.5 < d < 6$ nm in Eq. (4). Therefore, a good agreement is obtained with the effective mass theory assuming infinite potential barriers [34]. The thickness variation of the absorption edge energy shown in **Figure 9** demonstrates a remarkable blue shift of the spectra as the Si layer thickness decreases. This shift can only be caused by the Si layer because the SiO_2 barrier layer thickness is a constant value of 4.8 nm. This absorption edge energy is in accordance with E_0 (eV) = 1.61 + 0.75 d^{-2} (eV).

$$E_0 = Eg + \frac{\hbar^2 \pi^2 n^2}{2 m_0 d^2}\left(\frac{1}{m_e} + \frac{1}{m_h}\right) = E_g + \frac{0.75}{d^2}\,(\text{eV})\quad (d{:}\text{nm}, m_e = 1, m_h = 1) \tag{4}$$

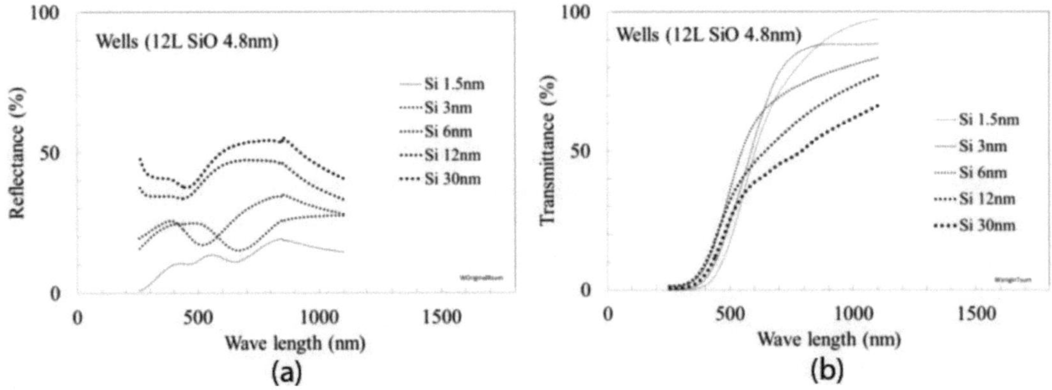

Figure 5. Optical properties of 12-layer Si/SiO_2 films of different QW thicknesses. The barrier thickness is 4.8 nm. (a) Optical reflectance spectra, (b) optical transmittance spectra.

http://dx.doi.org/10.5772/intechopen.79880

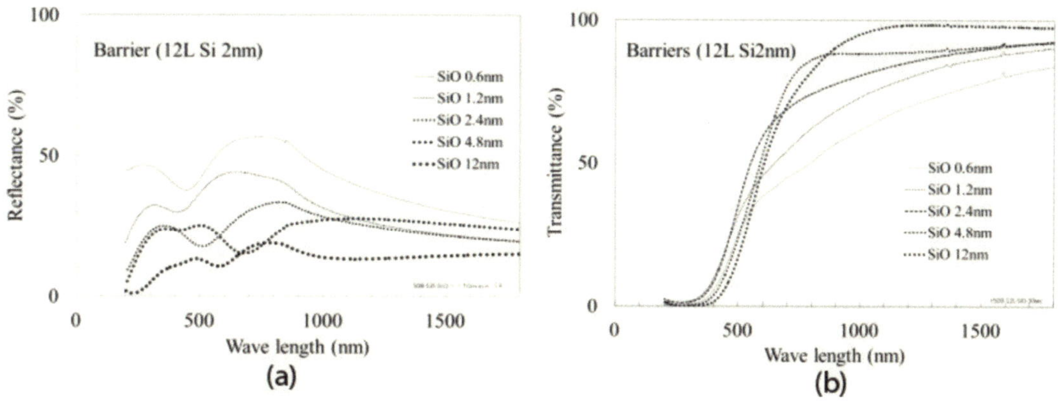

Figure 6. Optical properties of a 12-layer Si/SiO_2 film at a constant QW thickness of 2 nm for different barrier thicknesses. (a) Optical reflectance, (b) optical transmittance.

Although quantum confinement is obtained from the optical absorption measurements, the recombination mechanism is still indistinct. To elucidate the latter, we investigate PL spectra of the Si/SiO_2 multilayer nanostructures.

2.3. Thermal annealing of Si/SiO_2 layer films

Since the as-deposited samples show very weak photoluminescence, two experimental efforts are made to improve the PL intensity. The first is the increasing the well number of Si/SiO_2 films layered with 0.5–15 nm (Si) QWs from 10 to 50 periods. The second is the thermal annealing of Si/SiO_2 films in nitrogen. In our work, RTA in nitrogen was performed at 700 and 1100°C for 30 min. **Figure 10** shows the cross-sectional view of an RTA treated Si/SiO_2 film. Apparently, the Si QW layers are changed revealing partially dark spots and eroded SiO_2 barrier layers.

Figure 11 shows the XRD spectra of a 10-layer Si/SiO_2 film on a sapphire substrate which is rapid-thermal annealed at 700 and 1100°C. The crystallization is clearly identified at 700°C by the splitting the (111), (220), and (311) Braggs peaks indicating that the amorphous Si layers are crystallized as the nanocrystal Si.

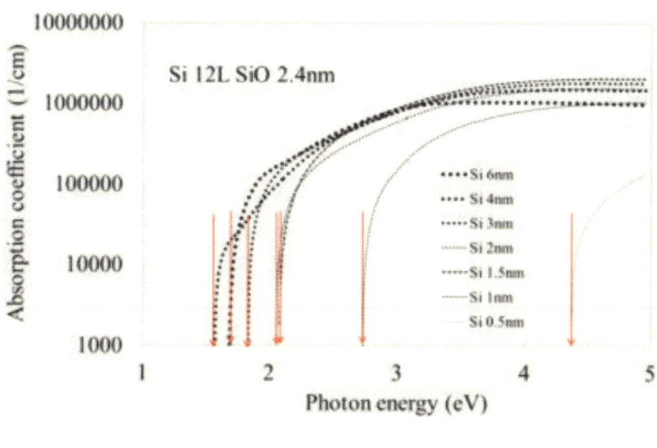

Figure 7. Absorption coefficient spectra for a 12-layer Si/SiO_2 film with a 2.4-nm SiO_2 barrier for different thicknesses of the Si well layer.

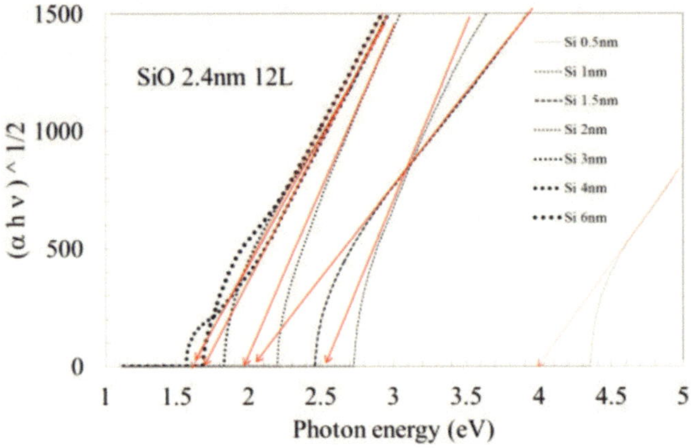

Figure 8. An enlarged view of absorption coefficient spectra displayed in **Figure 7**. The extrapolation gives the optical band gap energy.

2.3.1. Photoluminescence of Si/SiO_2 layer films

Photoluminescence spectra of as-deposited amorphous 10 layers Si/SiO_2 films are excited at 325 nm by a He-Cd laser. The highest energy peaks at 2.35, 2.05, 1.81 eV, with subpeak at 1.45 eV are observed. The improved PL is observed upon crystallization of Si after subjecting the 50-layer Si/SiO_2 multilayer nanostructures to RTA at 700 and 1100°C as shown in **Figure 12**. The spectra show a broadband peak and shoulders. The main peak energies are 1.62, 1.68, and 1.45 eV. In **Figure 13**, photoluminescence spectra of Si/SiO_2 QWs annealed at 1100°C for 30 min in nitrogen are shown for the QW thickness ranging from 1.2 to 2.5 nm. The intensity becomes higher for the thinner QWs.

Figure 13 presents the well thickness dependence of PL spectra taken on the 50-layer Si/SiO_2 structure upon RTA in nitrogen at 1100°C. The strongest PL is observed for the thinnest Si QW (1.2 nm), fading as the QW thickness increases. **Figure 14** displays the temperature dependences of photoluminescence spectra. Among the three temperatures, the 80 K spectrum is the

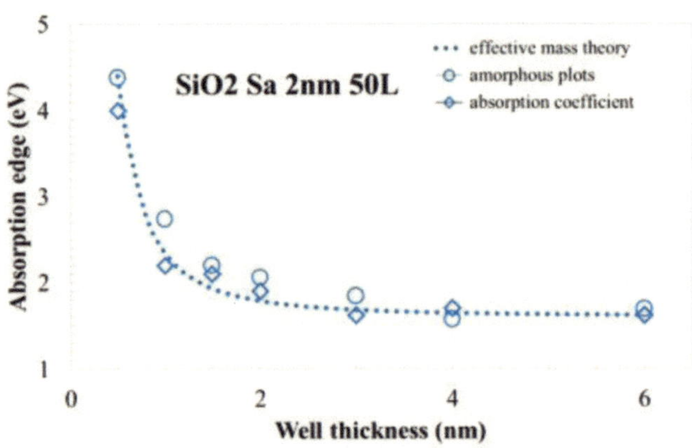

Figure 9. The absorption edge energy as a function of the QW thickness at the constant 2.4-nm SiO_2 barrier thickness in a Si/SiO_2 multilayer structure.

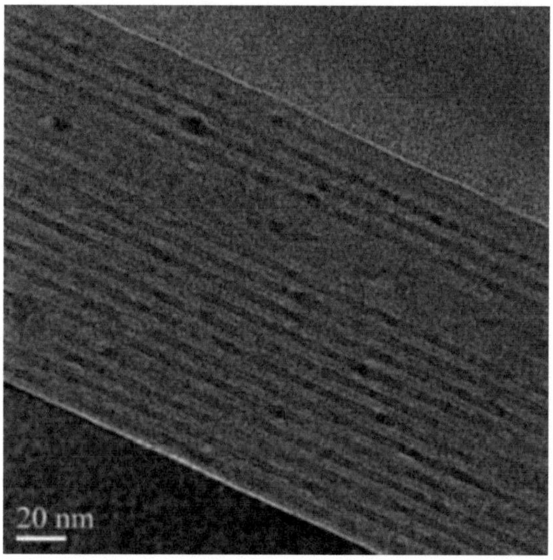

Figure 10. A cross-sectional view of SEM on an amorphous Si/SiO_2 film rapid thermal annealed at 1100°C in nitrogen.

most intense, followed by the room temperature, and the 4 K photoluminescence. The Kapoor model, where two different recombination mechanisms are operative in different temperature ranges, can explain this nonmonotonous temperature dependence of PL. The Kapoor's Eq. (5) consists of an Arrhenius-type term T_r and a Berthelot term T_B. Rolver explained the effects as an interplay of thermal activation of excitons into optically active states and hopping occupation of dark states.

$$I(T) = \frac{I_0}{1 + v_o \exp\left[\left(\frac{T}{T_B} + \frac{T_r}{T}\right)\right]} \quad (5)$$

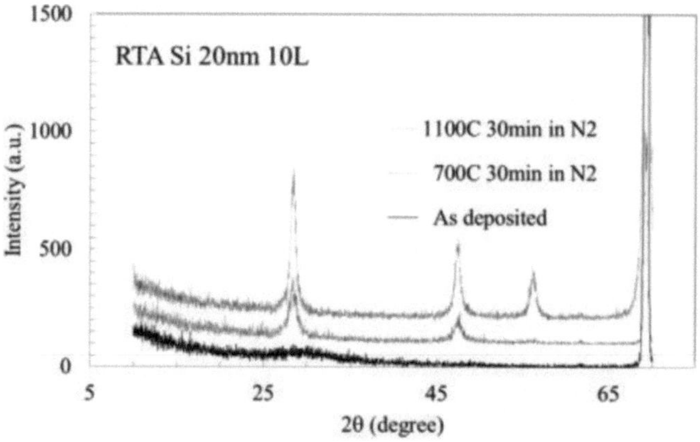

Figure 11. XRD spectra of a Si/SiO_2 multilayer nanostructure grown on a sapphire substrate and rapid-thermal annealed at 700 or 1100°C.

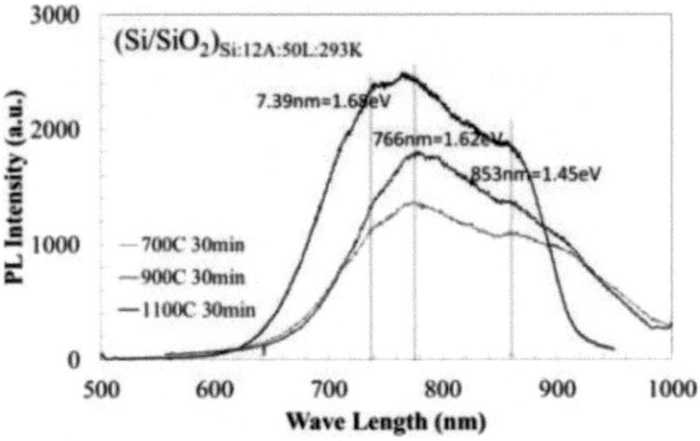

Figure 12. PL spectra of Si/SiO_2 QWs and annealing effects on the photoluminescence intensity of the 50-layer Si/SiO_2 (1.2 nm Si and 2.4 nm SiO_2) structures at room temperature.

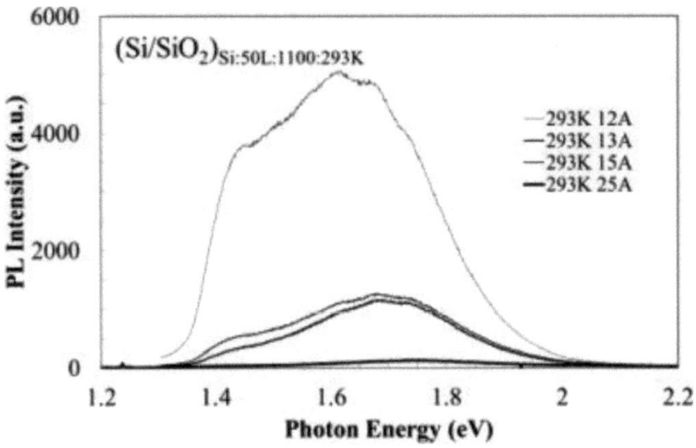

Figure 13. PL spectra of the 50 L Si/SiO_2 of different QW thicknesses (1.2, 1.3, 1.5, and 2.5 nm) annealed at 1100°C.

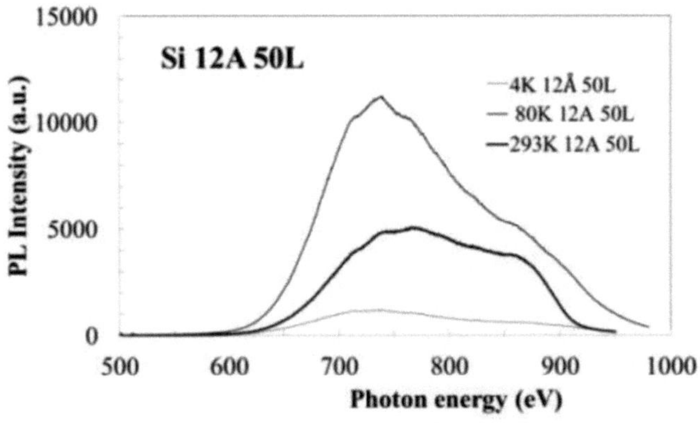

Figure 14. Temperature-dependent PL spectra. Intensity increases from 4 to 293 K and 80 K.

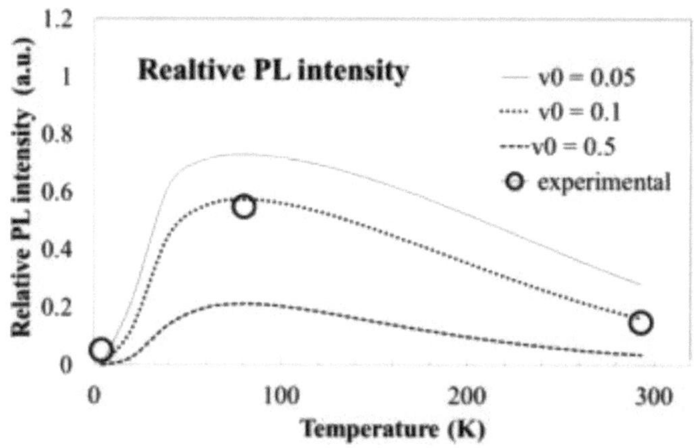

Figure 15. Temperature dependences of the PL peak intensity following the Kapoor model. The experimental results are consistent with the theory.

2.3.2. The Kapoor model

The temperature dependence of the photoluminescence intensity peaks observed at 80, 4, and 293 K are analyzed using the Kapoor empirical models [45, 67]. The simulation of the Si/SiO_2 sample comprising 50 quantum wells (1.2 nm well width) annealed at 1100°C for 30 min in nitrogen is performed following Eq. (5). **Figure 15** presents the result, which evidences a reasonable agreement between the experimental and simulated results using $T_r = 70$ K, $T_B = 80$ K, and $\upsilon_0 = 0.1$.

3. Summary

Amorphous nanostructured Si/SiO_2 films are smartly fabricated using a UHV RF magnetron sputtering system at room temperature. Absorption coefficients are evaluated considering the tentative well Si thickness and energy band gap energy of the Si/SiO_2 layers. The photon energy dependence of absorption coefficient on the quantum well thickness is simulated taking into account the quantum-confined properties. The choice of the Si layer thicknesses interfacing the SiO_2 barrier layer of the constant thickness (4.8 nm) mainly determines the blue shift of the absorption energy. Assuming the infinite potential SiO_2 barriers, the effective-mass theory provides the fitted absorption coefficient edge energy in accordance with E (eV) = 1.61 + 0.75 d^{-2} (eV) for one-dimensionally confined amorphous Si (d: nm). The amorphous Si/SiO_2 nanostructure films show the quantum confinement. Thermal annealing of the Si/SiO_2 films affects the improvement of photoluminescence intensity. Anomalous temperature dependence of photoluminescence is attempted to be explained based on the Kapoor model. Future work is expected to resolve many more research questions.

Acknowledgements

The authors express deep thanks to Mr. Minoru Kondo, a graduate student of Waseda University, for corporate experimental works and to Dr. Atsushi Kawaharatsuka, Associate Professor of

Waseda University, for many impressive discussions on quantum confinement. One of the authors (T. T.) expresses deep thanks to Dr. Tadashi Takahashi, Emeritus Professor at Tohoku University, for his encouragements and instructions.

Author details

Toshio Takeuchi[1]* and Yoshiji Horikoshi[2]

*Address all correspondence to: toshio-takeuchi@ve.cat-v.ne.jp

1 Sendai National College of Technology, Sendai, Japan

2 School of Science and Engineering, Waseda University, Tokyo, Japan

References

[1] Ley L, Kowalczyk S, Pollak R, Shirley DA. X-ray photoemission spectra of crystalline and amorphous si and ge valence bands. Physical Review Letters. 1972;**29**:1088-1092

[2] Abeles B, Tiedje T. Amorphous semiconductor superlattices. Physical Review Letters. 1983;**51**:2003-2006

[3] Tiedje T, Abeles B, Brooks BG. Energy transport and size effects in the photoluminescence of amorphous-germanium/amorphous-silicon multilayer structures. Physical Review Letters. 1985;**54**:2545-2548

[4] Wilson BA, Taylor CM, Harbison CM. Photoluminescence in ultrathin a-Si:H layers. Physical Review B. 1986;**34**:8733-8739

[5] Yang L, Abeles B. Quantum shift of the optical absorption edge in ultrathin amorphous hydrogenated germanium. Applied Physics Letters. 1987;**51**:264-266

[6] Miyazaki S, Ihara Y, Hirose M. Resonant tunnelingthrough amorphous silicon-silicon nitride double barrier structures. Physical Review Letters. 1987;**59**:125-127

[7] Kalm S. Optical investigation of a-Si:H/a-SiNx:H superlattice. Physical Review B. 1988;**37**: 8837-8841

[8] Hattori K, Mori T, Okamoto H, Hamakawa Y. Optical observation of subbands in amorphous silicon ultrathin single layers. Applied Physics Letters. 1988;**53**:2170-2172

[9] Yang L, Abeles B, Eberhardt W, Stasiewski H, Sondericker D. Photoemission spectroscopy of heterojunctions of hydrogenated amorphous silicon with silicon oxide and nitride. Physical Review B. 1989;**39**:3801-3815

[10] Lockwood DJ. Optical properties of porous silicon. Solid State Communications. 1994;**92**: 101-112

[11] Kovalev D, Pollisski G, Ben-Chorin MM, Diener J, Koch F. The temperature dependence of the absorption coefficient of porous silicon. Journal of Applied Physics. 1996;**89**:5978-5983

[12] Mulloni V, Chierchia R, Mazzoleni C, Pucker G, Pavesi L. Porous silicon optical devices and Si/SiO_2 quantum wells :recent results. Philosophical Magazine. 2000;**80**:705-718

[13] Kanemitsu Y, Okamoto S. Visible luminescence from silicon quantum dots and wells. Materials Science and Engineering. 1997;**B48**:108-115

[14] Kanemitsu Y, Fukunishi Y, Iiboshi M, Okamoto S, Kushida T. Visible luminescence from Si/SiO_2 quantum wells and quantum dots: confinement and localization of excitons. Physics E. 2000;**7**:456-460

[15] Moskalenko AS, Berakdar J, Prokofiev AA, Yassievch IN. Single-particle states in spherical Si/SiO_2 quantum dots. Physics Review. 2007;**76**:08547-1~7

[16] Aberle AG, Glunz SW, Stephens AW, Green MA. High efficiency silicon solar cells: Si/SiO_2 interface parameters and their impact on device performance. Progress in Photovoltaic Research and Application. 1994;**2**:265-273

[17] Kim DH. Current-voltage characterization of silicon quantum dot solar cells. Transactions on Electrical and Electronic Materials. 2009;**10**:143-145

[18] Filikhin I, Matinyan SG, Schmid BK, Vlahovic B. Electronic and level statistics properties of Si/SiO_2 quantum dots. Physics E. 2010;**42**:1979-1983

[19] A-Ameer NM, Abdulrida MC. Direct optical energy gap in amorphous silicon quantum dots. International Journal of Modern Physics. 2011;**2**:1530-1537

[20] Okamoto S, Kanemitsu Y. Quantum confinement and interface effects on photoluminescence from silicon single quantum wells. Solid State Communications. 1997;**103**:573-576

[21] Kanemitsu Y. Excitons in silicon quantum structures. Journal of Luminescence. 1999;**83-84**: 283-290

[22] Kanemitsu Y, Iiboshi M, Kushida T. Photoluminescence spectrum of a-Si/SiO_2 and c-Si/SiO_2 quantum wells. Journal of Luminescence. 2000;**87-89**:463-465

[23] Kanemitsu Y, Iiboshi M, Kushida T. Photoluminescence dynamics of amorphous Si/SiO_2 quantum wells. Applied Physics Letters. 2000;**76**:2200-2202

[24] Kanemitsu Y. Efficient light emission from crystalline and amorphous silicon nanostructures. Journal of Luminescence. 2002;**100**:209-217

[25] Guha S, Qadri SB, Musket RG, Wall MA, Shimizu-Iwayama T. Characterization of Si nanocrystalline grown by annealing SiO_2 films with uniform concentrations of implanted Si. Journal of Applied Physics. 2000;**88**:3954-3961

[26] Lu ZH, Grozea D. Crystalline Si/SiO_2 quantum wells. Applied Physics Letters. 2002; **80**:255-257

[27] Lockwood DJ, Lu ZH, Grozea D. Photoluminescence in crystalline-Si/SiO_2 quantum wells. Proceeding of SPIE. 2002:**4808**:40-44

[28] Mangolini I, Thimsen E, Kortshagen U. High-Yield Plasma Synthesis of Luminescent Silicon Nanocrystals. Nano Letters. 2005;**5**:655-659

[29] Schubert EF, Vredenberg AM, Hunt NE, Wong YH, Becker PC, Poate JM, Jacobson DC, Feldman LC, Zydzik GJ. Giant enhancement of luminescence intensity in Er-doped Si/ SiO_2 resonant cavities. Applied Physics Letters. 1992;**61**:1381-1383

[30] Watanabe K, Fujii M, Hayashi S. Resonant excitation of Er3+ by the energy transfer from Si nanocrystals. Journal of Applied Physics. 2001;**90**:4761-4767

[31] Gourbilleau F, Dufour C, Madelon R, Rizk R. Spectroscopic studies of Er-doped Si-rich oxide/SiO_2 multilayers. Optica Applicata. 2007;**37**:21-29

[32] Lu ZH, Lockwood DJ, Barbeau JM. Quantum confinement and light emission in SiO_2/Si superlattices. Nature. 1995;**378**:258-260

[33] Vervoot L, Bassani F, Mihalceescu I, Vial JC, Davitani FA. Efficient visible-light emission from Si/SiCaF2(111) heterostructures grown by molecular beam epitaxy. Physica Status Solidi. 1955;**190**:123-127

[34] Lockwood DJ, Lu ZH, Barbeau JM. Quantum confinement luminescence in Si/SiO_2 superlattices . Physical Review Letters. 1996;**76**:539-541

[35] Lu ZH, Lockwood DJ, Barbeau JM. Visible light emitting Si/SiO_2 superlattices. Solid-State Electronics. 1996;**40**:197-201

[36] Tsu R, Filios A, Lofgren C, Cahill D, Vannostrand J, Wang CG. An epitaxial Si/SiO_2 superlattice barrier. Solid State Electronics. 1996;**40**:221-223

[37] Khriachtchev L, Kilpela O, Karirinne S, Keranen J, Lepisto T. Substrate-dependent crystallization and enhancement of visible photoluminescence in thermal annealing of Si/SiO_2 superlattices. Applied Physics Letters. 2001;**78**:323-325

[38] Nassiopoulou AG, Ioannou-Sougleridis V, Photopoulos P, Travlos P, Tsakiri V, Paradimitriou D. Stable visible photo- and electroluminescence from nanocrystalline silicon thin films fabricated on thin SiO_2 layers by low pressure chemical vapor deposition. Physica Status Solidi, (a). 1998;**165**:79-85

[39] Heikkila L, Kuusela T, Hedman HP, Ihantola H. Electroluminescent SiO_2/Si superlattices prepared by low pressure chemical vapor deposition. Applied Surface Science. 1998;**133**:84-88

[40] Tsybeskov L, Hirchman KD, Duttaguta SP, Zacharias M, Fauchet PM, McCaffrey JP, Lockwood DJ. Nanocrystalline-silicon superlattice produced by controlled recrystallization. Applied Physics Letters. 1998;**72**:43-45

[41] Vinciguerra V, Franzo G, Priolo F, Lacona F, Spinella C. Quantum confinement and recombination dynamics in silicon nanocrystals embedded in Si/SiO_2 superlattices. Journal of Applied Physics. 2000;**87**:8165-8172

[42] Park N, Choi C, Seong T, Park S. Quantum confinement in amorphous silicon quantum dots embedded in silicon nitride. Physical Review Letters. 2001;**86**:1355-1357

[43] Arguirov T, Mchedlidze T, Arguirova SK, Kitter M. Light induced solid-phase crystallization of Si nanocrystals in Si/SiO_2 multiple quantum wells. Journal of Applied Physics. 2002;**107**:124302-124309

[44] Wang XX, Zhang JG, Ding L, Cheng BW, Ge WK, Yu JZ, Wang QM. Origin and evolution of photoluminescence from Si nanocrystals embedded in a SiO_2 matrix. Physical Review B. 2005;**72**:195313-1~6

[45] Rolver KR, Fost M, Winkler O, Spangenberg B, Kurz H. Influence of excitonic singlet-triplet splitting on the photoluminescence of Si/SiO_2 multiple quantum wells fabricated by remote plasma-enhanced chemical-vapor deposition. Journal of Vacuum Science and Technology A. 2006;**24**:141-145

[46] Arguirov T, Mchedlidze T, Akhmetov VD, Arguirova SK, Kitter M, Rover R. Berghoff B, Forst M, Batzner DL, Spangenberg S. Effect of laser annealing on crystallinity of the Si layers in Si/SiO_2 multiple quantum wells. Applied Surface Science. 2007;**254**:1083-1086

[47] Bergfoff B, Forst M, Batzner DL, Spangenberg S. Electronic band gap of Si/SiO_2 quantum wells: Comparison of ab initio calculation and photoluminescence measurements. Applied Surface Science. 2007;**254**:1083-1086

[48] Arguirov T, Mchedlidze T, Arguirova SK, Kitter M, Rover R, Berghoff B, Batzner D, Spangenberg S. Laser annealing of the Si layers in Si/SiO_2 multiple quantum wells. Materials Science and Engineering B. 2009;**159-160**:57-60

[49] Puglisi RA, Vecchio C, Lombardo S, Lorenti S, Camallen MC. Charge transport in ultra-thin silicon rich oxide/SiO_2 multilayers under solar light illumination and in dark conditions. Journal of Applied Physics. 2010;**108**:023701-1~7

[50] Kanzawa Y et al. Size-dependent near-infrared photoluminescence spectra of Si nanocrystals embedded in SiO_2 matrices. Solid State Communications. 1979;**102**:533-537

[51] Zayats AV, Yu A, Repeyev A, Nikogosyan DN, Vinogrado EA. Radiative recombination in short-period a-Si/SiO_2 superlattices. Journal of Luminescence. 1972;**52**:335-343

[52] Dobrowolski JA, Pekelsky JR, Pelletier R, Ranger M, Sullivan B, Waldolf AJ. Practical magnetron sputtering system for the deposition of optical multilayer coating. Applied Optics. 1992;**31**:3784-3789

[53] Sullivan B, Dobrowolski JA. Applied Optics. Deposition error compensation for optical multilayer coating. II. Experimental Results-Sputtering System. 1993;**32**:2351-2360

[54] Sullivan B, Byrt KL. Metal/dielectric transmission interference filters with low reflectance. 2. Experimental results. Applied Optics. 1995;**34**:5684-5694

[55] Zacharias M, Tsybeskov L, Hirschman KD, Fauchet PM, Blasing J, Veit P, Blasting J, Kohlert P, Veit P. Nanocrystalline silicon superlattices: Fabrication and characterization. Journal of Non-Crystalline Solids. 1998;**227-230**:1132-1136

[56] Qin GG, Ma SY, Ma ZC, Zong WH, Liping Y. Electroluminescence from amorphous Si/SiO_2 superlattices. Solid State Communications. 1998;**106**:329-333

[57] Zacharias M, Blasing J, Veit P, Tsybeskov L, Hirschman K, Fauchet PM. Thermal crystallization of amorphous Si/SiO_2 superlattices. Applied Physics Letters. 1999;**74**:2614-2616

[58] Zacharias M, Streitenberger P. Crystallization of amorphous superlattice in the limit of ultrathin films with oxide interfaces. Physical Review B. 2000;**62**:8391-8396

[59] Benyoucef M, Kuball M, Sun JM, Zhong GZ, Fan XW. Raman scattering and photoluminescence studies on Si/SiO_2 superlattices. Journal of Applied Physics. 2001;**89**:7903-7907

[60] Cho EC, Xia J, Aberle AG, Green MA. Antireflection and surface passivation behavior of SiO_2/Si/SiO_2 quantum wells on Si. Solar Energy Materials & Solar Cells. 2002;**74**:147-154

[61] Tenron C, Gourbilleau F, Rizk R, Dufour C. Si/SiO_2 multilayers: synthesis by reactive magnetron sputtering and photoluminescence emission. Physica E. 2003;**16**:517-522

[62] Hanaizumi O, Ono K, Ogawa Y. Blue-light emission from sputtered Si:SiO_2 films without annealing. Applied Physics Letters. 2003;**82**:538-540

[63] Portier X, Ternon C, Gourbilleau F, Dufour C, Rizk R. Anneal temperature dependence of Si/SiO_2 superlattices photoluminescence. Physica E. 2003;**16**:439-444

[64] Huang S, Xiao H, Shou S. Annealing temperature dependence of Raman scattering in Si/SiO_2 superlattice prepared by magnetron sputtering. Applied Surface Science. 2009;**255**:4547-4550

[65] Takeuchi T, Kondo M, Fujita M, Kawaharazuka A, Horikoshi Y. Optical properties of amorphous and nanostructured Si/SiO_2 quantum wells. Journal of Nano Research. 2014; **26**:59-62

[66] Keister JW, Rowe JE, et al. Structure of ultrathin SiO_2/Si(111) interface studied by photoelectron spectroscopy. Journal of Vacuum Science and Technology. 1999;**A17**:1250-1257

[67] Kapoor M, Sinh VA, Johri GK. Origin of the anomalous temperature dependence of luminescence in semiconductor nanocrystalline. Physical Review B. 2000;**61**:1941-1945

Chapter 8

Self-Assembled Nanostructures Formation in Hybrid Metal-Mesogenic Systems

Tatyana I. Shabatina

Additional information is available at the end of the chapter

http://dx.doi.org/10.5772/intechopen.83003

Abstract

This chapter reviews the methods of supramolecular chemistry and cryochemistry to study the formation, morphological and structural properties, and possible applications of hybrid nanostructures and nanosized aggregates comprising plasmonic metals (silver and copper) and several mesogenic compounds, which exhibit different liquid crystalline mesophases—nematics, smectics, and cholesterics. Low-temperature vacuum co-condensation of reagent vapors on cooled surfaces of quartz, KBr, CaF_2, or polished copper was used to synthesize hybrid nanosystems including silver and copper and long-chain mesogenic derivatives of alkylcyanobiphenyls under molecular beam conditions. Controlled thermal treatment of the samples allowed the directed formation of metal nanoparticles of definite size from 2 up to 100 nm. It was shown that the procedures of temperature treatments and molecular self-organization of different liquid crystalline phases controlled the size and morphology of nanoparticles and their aggregates, which were formed in the system. Transmission electron microscopy (TEM) and atomic force microscopy (AFM) data of the samples show the formation of orientationally ordered structures in nematic mesophases. Formation of flat 2D aggregates was found in layered smectic mesophases. Optical absorbance spectra of silver/4-pentyl-4-cyanobiphenyl (Ag/5CB) and copper/4-pentyl-4-cyanobiphenyl (Cu/5CB) samples encapsulated in polymer poly-para-xylylene at 300 K contained characteristic bands of plasmonic absorbance of metal nanoparticles at 420–440 nm and 560–600 nm. Rising metal concentration in the sample led to the performable growth of rod-like metal particles with anisometric ratio $l/d > 20$ and intensive absorbance at higher wavelengths ($\lambda \geq 650$ nm). New hybrid nanosystems based on biomolecules of cholesterol or its heteroatomic analog thiocholesterol including nanosized silver particles of $d = 5.0 \pm 0.5$ and $d = 2.5 \pm 0.5$ nm, respectively, were obtained. Highly ordered 1D-, 2D-, and 3D structures containing silver nanoparticles were formed from concentrated organic sols at different support surfaces by removing the inert solvent from triple metal/ligand/solvent system and by cooling the binary metal/ligand system from isotropic state to the cholesteric liquid crystalline mesophase. The microstructure and composition of the hybrid nanosystems were characterized by FTIR, UV-Vis spectroscopy, TEM, and selected area electron diffraction (SAED). It was shown that hybrid nanosystems based on silver nanoparticles covered by a stabilizing layer of mesogenic molecules of thiocholesterol display selectivity in adsorption of optical isomers of the selected compounds.

Keywords: hybrid nanostructures, metal-mesogenic nanosystems, nanosized metal particles, self-assembling, nanoparticle's super-lattices, optical properties, adsorptive properties

1. Introduction

This chapter reviews the methods of supramolecular chemistry and cryochemistry applied to study hybrid nanostructures and nanosized aggregates formed on plasmonic metals (silver and copper) and some mesogenic (liquid crystalline) compounds, which, in certain temperature ranges, reveal various liquid crystalline mesophases—orientationally ordered nematics, smectics with translationally ordered layer structure, and cholesterics exhibiting the space helix structures due to the chirality of their molecular structure and specifics of their helix supramolecular ordering. Attention was also paid to the peculiarities of morphological and structural behavior of such hybrid nanosystems, of their optical and adsorptive properties as well as perspectives of their possible applications.

Hybrid nanosystems comprising nanosized metal particles stabilized in functionalized organic matrices are scientifically interesting due to their highly promising applications in nanoelectronics, photonics, and sensor systems [1–6]. The use of mesogenic compounds in different liquid crystalline states as soft ordered template matrices opens new possibilities for the synthesis of anisotropic metal particles and their highly ordered assemblies. The presence of metal atoms and ions in liquid crystalline systems significantly extended the possible practical areas of their exploitation. The most prospective are hybrid metal-mesogenic nanosystems, combining molecular ordering of mesogenic matrix compounds and quantum-size properties of nanometals, which are different from both the characteristics of the single atoms and the properties of bulk materials. The properties of such systems are dependent not only on the size and shape of individual nanoparticles of the formed system, but also on the character of their joint organization and formation of the hybrid ordered structure [7, 8]. Among different methods of metal particles' incorporation into the organic systems and formation of hybrid nanosystems, the most original and useful are methods of low-temperature condensation of metal atoms with the molecular beam of organic component's vapors under high-vacuum conditions [9, 10]. The use of low temperatures and technique of vacuum condensation of metal atoms and organic molecular beams on the cooled support surfaces allowed researchers to stabilize and study such highly active species as metal clusters and nanoparticles and their labile complexes. This approach allowed us to realize the formation of hybrid nanosystems by bottom-up method and to carry out the directed metal atom and small clusters aggregation, controlled by structural and dynamic properties of molecular organized matrices, and also to create new ordered assemblies of nanoparticles.

In this work, the formation of new hybrid metal-mesogenic nanosystems was exemplified by two d-metals of the 11th group—silver and copper with mesogenic derivatives of long-chain alkylcyanobiphenyls. Nanoparticles of silver and copper possess the intensive plasmonic bands in the visible region of spectrum and can be used for the visualization and marking of different biological objects. Mesogenic compounds of the alkylcyanobiphenyl row are widely used nowadays as components of active layers for liquid crystalline displays and indicators. These compounds form ordered liquid crystalline phases of two types—nematic, characterized by the existence of molecular orientations of long-range order, and smectic—with layered molecular ordering. The existence of molecular associates and formation of dimeric structures in solid and liquid crystalline phases is known for these compounds at low temperatures. They form several metastable and stable solid phases with a local structure similar to the liquid crystalline mesophases with a different molecular organization [11, 12]. A terminal cyano group and an aromatic core of the molecules can be considered as ligands suitable for the formation of complexes of π- and σ-types with atoms and clusters of the metals under consideration [13].

Controlled formation of anisometric nanoparticles and ordered hybrid nanostructures is one of the main objectives of this work. The role and mutual influence of self-organization processes of metal species and organic component's molecules were considered. The phase-structural transitions and physico-chemical properties that appear for new hybrid metal-mesogenic nanosystems were demonstrated on the example of a combination of silver and copper metal nanoparticles formed in mesogenic matrices of long-chain ($n \geq 5$) alkylcyanobiphenyls 4-pentyl-4-cyanobiphenyl (5CB) and 4-octyl-4-cyanobiphenyl (8CB).

Mesogenic derivatives of cholesterol possess unique liquid crystalline mesophases named cholesteric mesophases at definite temperature ranges [1]. The molecules in such a mesophase are arranged in prolonged 3D helix structures with special optical properties depending on the pitch of the helix, such as selective reflection of visible light. Introduction of metal atoms and clusters in cholesteric mesophase changes the intermolecular interactions in the system and leads to the induction of new liquid crystalline and isotropic phases. We can also find the formation of self-assembled nanosized aggregates with different morphology and hybrid nanostructures, including metal species of different sizes arranged by a mesogenic matrix in highly ordered assemblies with modified optical properties. Special attention will be paid to analyzing the possibilities of their use in modulation of optical signals and obtaining systems with tunable optical plasmonic resonance.

Self-organization of nanosized metal particles in highly ordered assemblies with definite structural and morphological characteristics is nowadays one of the most promising areas of nanoscience and nanotechnology [2, 3]. The unique quantum-size properties and the appearance of new collective modes open new possibilities for their application in such fields as microelectronics, optics, catalysis, information recording systems, and data storage systems [4, 5, 7]. The physical and chemical properties of such nanohybrid systems depend both on the size and shape of the structural elements (nanoparticles), and on the manner of their organization into the joint structure [8, 9]. Development of new synthesis methods of nanohybrid aggregates possessing special order in regular placement of the particles of the same size is an important task in modern nanochemistry. For the synthesis of nanoparticles having a certain size, it is necessary to physically or chemically isolate nanoparticles from one another and from the external environment to prevent aggregation. In many cases, the stability of such systems is achieved through the introduction of functionalized organic ligands aryl- and alkylthiols, organic amines, alcohols, and other compounds capable of specific interactions with the surface of the nanoparticles [10, 11]. Hybrid nanosystems including biomolecules or their chemical analogs attract the increasing interest of researchers due to their compatibility with natural systems [12–14]. In the present chapter, we show the possibilities of using two new silver-containing nanosystems based on biomolecule of cholesterol and its heteroatomic analog—thiocholesterol. Supramolecular organization of the ligand molecules allows us to use the obtained nanoparticles for self-assembling in binary superlattices and for the formation of ordered 1D-, 2D-, and 3D nanostructures.

Cholesterol and its derivatives are optically active molecules and form helical liquid crystalline structures—cholesteric mesophases—at definite conditions. It is known that the effects of "guest-host" interactions are very important for the practical applications of composite liquid crystalline systems, including guest molecules of organic and inorganic compounds [15, 16]. At the same time, these compounds contain such functional groups as –OH and –SH, which are able to the effective interaction with the surface atoms of metal nanoparticles and form a mesogenic ligand stabilizing and selecting layers on the surface of metal nanoparticles. Recently, we have intensively developed investigations on the creation of new methods of producing hybrid metal-mesogenic nanosystems and the study of their physical and chemical properties [17–20].

2. Morphology of metal nanoparticles, formed in alkylcyanobiphenyl matrices at low temperatures

TEM study of the film samples of hybrid nanosystems "silver-4-pentyl-4-cyanobiphenyl" (Ag/5CB) and "copper-4-pentyl-4-cyanobiphenyl" (Cu/5CB), obtained by low-temperature co-deposition of components and annealing at 150–200 K for 1–2 h, shows the possibility of the formation and stabilization of mesogenic matrix by 5CB practically monodispersed spherical metal nanoparticles 1–2 nm in size (**Figure 1**). These metal species were formed in the solid nanostructured matrix of 5CB as a result of thermal decomposition of unstable dimeric biligand π-complexes of metal atoms $M(5CB)_2$ [16]. The metal atoms liberated by degradation of the complexes localize and can interact with each other in the local areas formed by mobile alkyl chains of a terminal substituent in 5CB dimeric structures existing at such low temperatures. Results of the study of different small areas of the film sample by selected area electron diffraction (SAED) show the formation of metal nanoparticles with fcc lattice. The size of metal nanoparticles formed in mesogenic matrices at these temperatures was close to the critical crystal size of the metal particles.

These small metal particles are stable at low temperatures under the conditions of limited molecular mobility of mesogenic matrix particles and can play a role of the initial crystalline particles (the embryo of a new phase) by raising the temperature. These small metal crystals appear in ESR spectra of the samples as broadened singlets with g-factor close to the value for free electron, g_e = 2003, which could be referred to conducting electrons spin resonance in metal nanosized species [16].

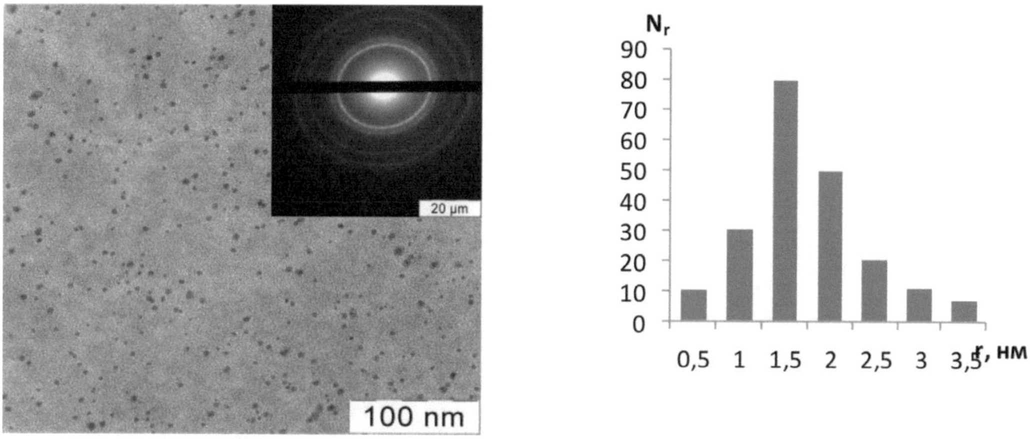

Figure 1. TEM pictures and electronic diffraction of the samples, including silver nanoparticles stabilized in mesogenic 5CB matrix at 200 K and histogram of their size distribution.

By annealing the samples from 90 up to 150 K, the relative intensity of the central component in ESR spectra increases, which reveals the thermal decomposition of the metastable complexes and the simultaneous growth of metal nanoclusters. UV-Vis spectroscopic study of silver/CB and copper/CB systems showed the appearance of new plasmonic absorbance bands in the visible region with a maximum at 400 and 560 nm, respectively.

On further heating of the samples up to temperatures more than 200 K, matrix crystallization process took place, which led to a sharp increase of matrix molecular mobility and rapid coalescence of nanoclusters followed by forming bigger aggregates, which was not recorded by the ESR technique. Such thermal behavior of the hybrid system can be deduced from the specifics of self-assembling processes and supramolecular organization of mesogenic cyanobiphenyl matrix at different temperatures. The combination of rigid aromatic core and flexible hydrocarbon chains of terminal substituents in ligand molecules leads to the formation of dimeric cyanophenyl structures in solid and liquid crystalline states at low temperatures. Such supramolecular structures could stabilize metal atoms due to the formation of metastable complexes. Thermal degradation of the complexes by increasing the temperature of the co-condensate samples followed by metal atoms escape into more soft domains formed by hydrocarbon chains and their rapid aggregation. Stabilization of nanosized metal clusters of size 1–2 nm formed at this stage is possible due to their low diffusion mobility in the solid cyanophenyl matrix. Warming of co-condensate systems up to the temperatures more than 200 K ($T > 200$ K), corresponding to CB crystallization, causes matrix softening and rise in the translational mobility of matrix molecules. This process is followed by metal nanoclusters' coalescence and formation of higher nanoparticles. Rapid heating up of the co-condensates up to the room temperature followed by the phase transition of the system in nematic (orientationally ordered molecular long-axes) and smectic phases (molecular ordering layered structure) led to the performable formation of the anisometric nanoparticles. In metastable crystal state of 5CB at 273–293 K, the temperatures close to the crystal to nematic phase transition point for hybrid metal-mesogenic "Ag-5CB" nanosystem, under the conditions of maximal ordering of liquid crystalline matrix molecules, the directed growth of highly anisotropic metal

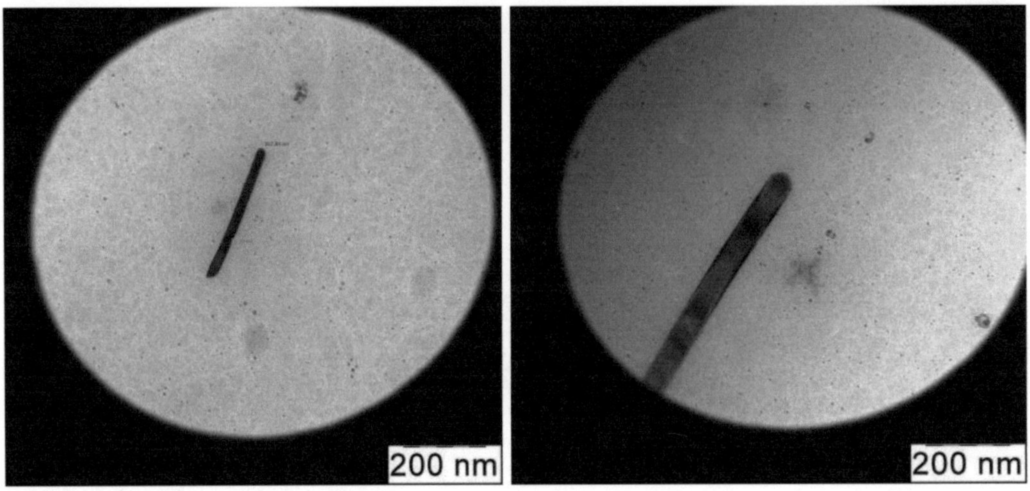

Figure 2. TEM pictures of silver nanorods, growing in 5CB matrix at 273–290 K.

nanorods took place, their lengths (l, nm) were more than 200 nm and their widths (d, nm) were about 15 nm, so anisometric l/d ratio was more than 10 (**Figure 2**). Mechanism of such almost single-dimensional growth of metal nanorods, in this case, is connected with the rapid growth of the molecular matrix mobility and maintaining high orientational order of the system.

Study of the samples by X-ray diffraction shows the exact crystal character of metal nanoparticles formed at these conditions. The samples possess texture that appeared due to the directed growth of rods along the crystallographic plane (100) and perpendicular to crystallographic plane (111). Observed mechanism of nanoparticles' growth can be explained by concurrent interaction of terminal functional cyano (CN) group of ligand molecules with metal atoms belonging to different crystallographic planes of forming nanoparticles. The interaction of ligand molecules with metal centers of highly packed bulky Ag (111) is slower in comparison with more loose Ag (100); so, the preferential growth of nanorods can be realized just in this direction due to the effective accession of small metal clusters and atom-by-atom addition formed in newly added layers and formation of highly anisometric metal nanoparticles.

3. Self-assembling of metal nanoparticles in nematic and smectic phases of mesogenic alkylcyanobiphenyls

It was shown by the combination of transmission electron microscopy (TEM) data with the results of selected area electron diffraction (SAED) and X-ray diffraction that hybrid metal-mesogenic nanosystems formed by low-temperature co-condensation of silver and copper with mesogenic cyanobiphenyls and followed by controlled annealing of the samples constitute of nanohybrid materials including metal nanoparticles stabilized and organized by matrix of mesogenic ligands. Depending on the phase of hybrid nanosystems Ag/5CB and

Ag/8CB, nanoparticles of different sizes and morphology—globular with main diameter 15 nm, anisometric rods, and flat 2D aggregates—were formed. Rapid heating of metal-mesogenic co-condensate samples up to 313 K was followed by phase transition of the hybrid system to the isotropic state and formation of globular metal particles of average diameter 15 nm. The data of FTIR study of the system show the formation of monomeric alkylcyanobiphenyl molecules stabilizing layer on the surface of nanoparticles due to donor-acceptor interactions of terminal CN group of ligand with surface metal atoms. The rising metal concentration and carrying out the nanoparticles' self-assembling processes in orientationally ordered nematic mesophase led to the formation of anisometric metal nanoparticles (anisometric factor l/d = 3–4) and their highly ordered assemblies. The results of AFM study of the samples show the formation of linear aggregates and regular orientationally ordered structures of anisometric nanoparticles in nematic phase 5CB at 300 K (**Figure 3**). Histogram of bimodal size distribution for metal nanoparticles in this case is also presented in **Figure 3**.

Using higher homology 4-octyl-4-cyanobiphenyl (8CB) molecules as stabilizing ligands gives us the possibility to carry out metal nanoparticles' formation in layered molecular structured smectic mesophase also. In this case, we saw the formation of quasi-fractal flat metal aggregates intercalated between smectic layers of 8CB due to higher interacting energy in the smectic layers and microphase segregation (**Figure 4**).

Thus, fulfilling the self-organization and self-assembly processes during the formation of hybrid nanosystems in liquid-crystalline mesophases of different types led to the production of metal nanoparticles of different sizes and morphology and their supramolecular organization in highly ordered nanostructures, including anisometric nanoparticles and rod-like aggregates stabilized in nematic mesophases and flat 2D metal aggregates in layered structured smectic phases of long-chain alkylcyanobiphenyls. Increasing the ligand molecules' concentration led to the stabilization of metal nanoparticles of smaller size. Increasing metal concentration in the orientationally ordered nematic system supported the formation of anisometric rod-like metal nanoparticles and their aggregates. The scheme of phase-structural states for Ag/

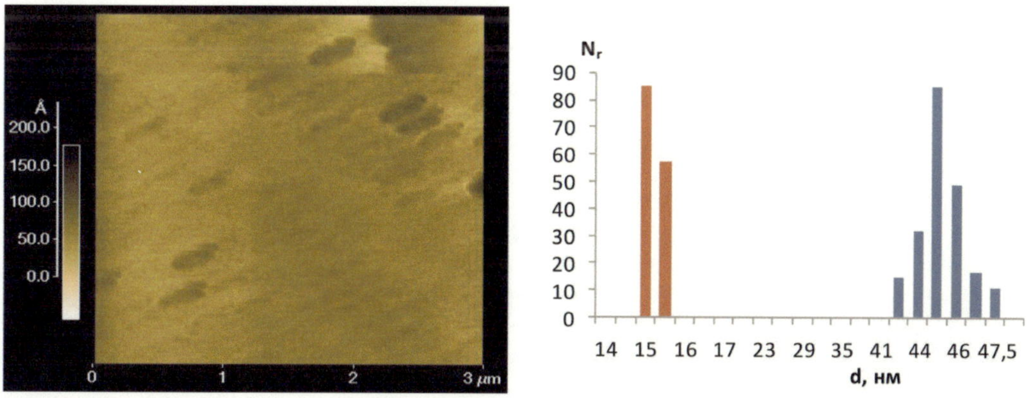

Figure 3. An AFM image (tapping mode) of the sample of hybrid metal-mesogenic nanosystem "Ag-5CB" (1/10) in nematic phase 5CB (left) and histogram of size distribution (right) for silver nanoparticles, red—cross size of nanoparticles, blue—linear length of nanoparticles.

5CB hybrid nanosystem indicating the conditions for the formation of metal nanoparticles with definite morphology is presented in **Figure 5**. This scheme is considered the combination of the result of thermal analysis, polythermal polarization spectroscopy, and electronic and atomic force microscopy.

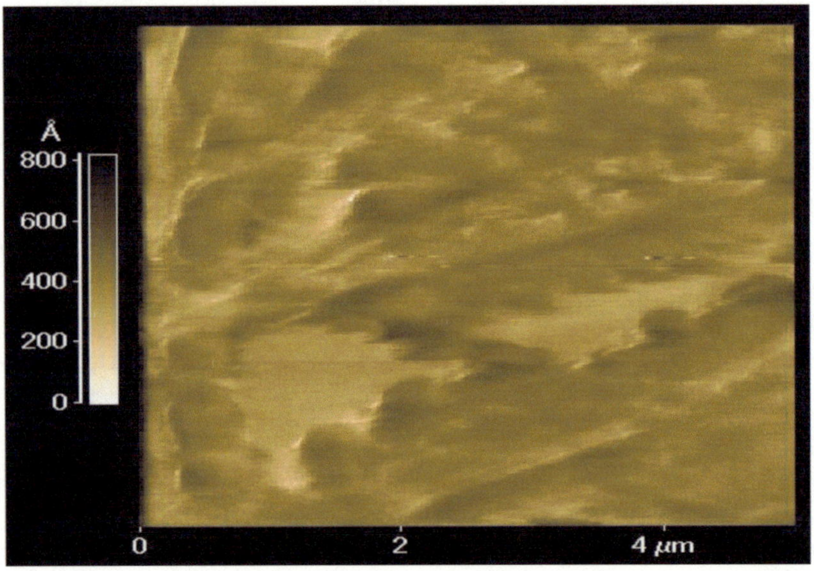

Figure 4. An AFM image (tapping mode) of the sample of hybrid metal-mesogenic nanosystem "Ag-8CB" (1/10) in smectic phase.

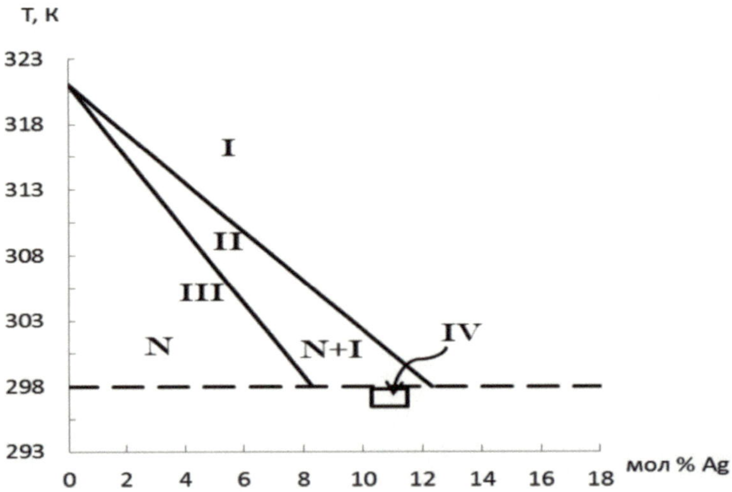

Figure 5. The scheme of phase-structural states for "silver-5CB" hybrid nanosystem: I—isotropic phase; II—two-phase region: isotropic phase including the fluctuations of nematic ordering; III—nematic phase; IV—the temperature region in nematic phase close to phase transition from crystalline to nematic; and V—solid phase at 150–200 K.

The data presented in **Figure 5** show the formation of nanoparticles with different morphology and sizes, stabilized by mesogenic ligands depending on the phase of mesogenic matrix and metal concentration in the system. The shape of the particles is in general the result of competition between general tendency for the minimization of particle's surface energy and elastic energy of liquid crystal. Rapid heating (more than 100 K/min) of the co-condensate Ag/5CB sample to temperatures of isotropic phase ($T > 318$ K)—region I—led to the formation and stabilization of global nanoparticles with $d = (15 \pm 10)$ nm. Transition to the nematic phase of 5CB gives us anisometric metal nanoparticles and their orientationally ordered assemblies—region III. In the bi-phase region II, including the co-existence of isotropic phase with fluctuations of nematic ordering, the formation of both isotropic globular and anisometric rods is possible. Maintaining the samples at 273–283 K and increasing the metal content in the sample from 1 up to 10 w/w % (region IV in **Figure 5**) led to the preferential growth of rod-like metal nanoparticles with l/d ratio more than 10. Small metal nanoparticles 1–2 nm in size are formed in diluted systems in the solid phase at 150–200 K.

4. Hybrid metal-mesogenic nanosystems encapsulated in polymer films and their optical properties

The layer-by-layer co-condensation of vapors of para-xylylene monomer, metal silver, and cyanobiphenyl ligand followed by controlled heating or UV irradiation of the sample led to the formation of hybrid metal-mesogenic compositions with varied metal contents from 1 up to 10 w/w %, encapsulated in polymer film:

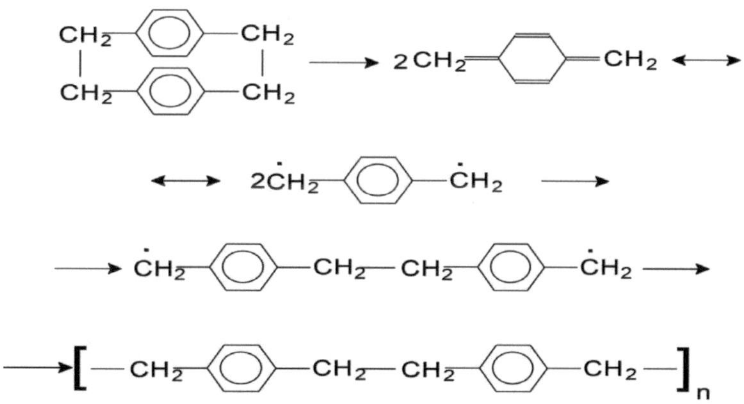

The results of TEM study of the Ag/5CB samples encapsulated in poly-para-xylylene (PPX) films show that increasing metal contents in the sample up to 10 w/w% led to the growth and stabilization of elongated rod-like metal particles with anisometric ratio l/d more than 10 (where l is the length and d is the cross diameter of particles). X-ray diffraction pictures of the Ag/5CB (w/w 1:10) sample, formed at 200 K followed by annealing up to 450 K, are shown in

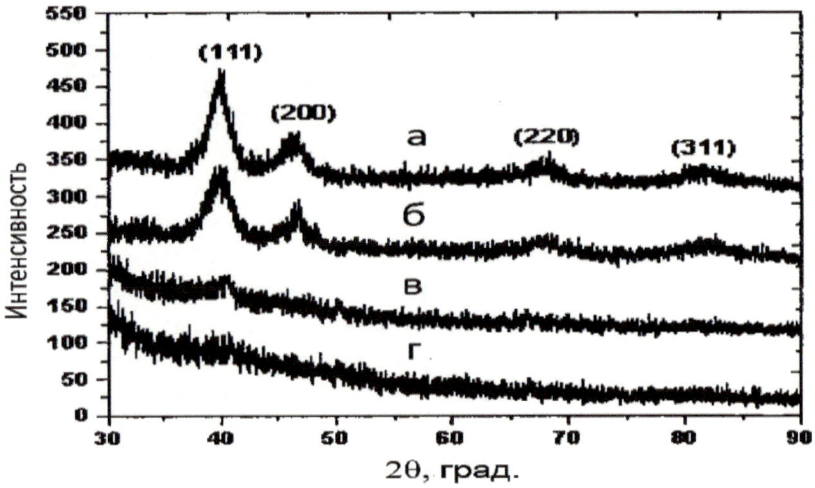

Figure 6. X-ray data for sample of hybrid metal-mesogenic nanosystem Ag/5CB (1:10), encapsulated in polymer (PPX) film at 450 K (a), 300 K (b), 250 K (c), and 200 K (d).

Figure 6. The results obtained show the formation of metal nanoparticles with fcc lattice, with lattice constant similar to that of crystal silver, a = 0.408 nm; the crystallographic planes were identified as (111), (200), (220), and (311). TEM data for the Ag/5CB (10 w/w %) sample encapsulated in polymeric film showed the formation of regular nanostructures stabilized by liquid crystalline matrix of anisometric metal nanoparticles (**Figure 7**), and their optical spectra are presented in **Figure 8**. Optical spectra contained the intensive absorbance bands of bimodal plasmonic resonance of anisometric silver nanoparticles (**Figure 8**).

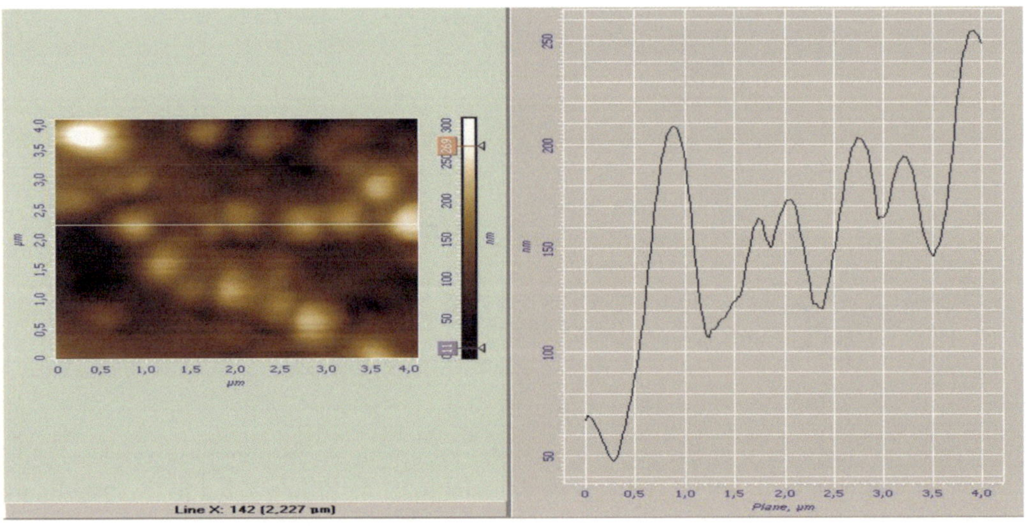

Figure 7. AFM data for the sample of hybrid metal-mesogenic nanosystem Ag/5CB (1:10 w/w%) encapsulated in poly-para-xylylene polymeric film.

http://dx.doi.org/10.5772/intechopen.83003

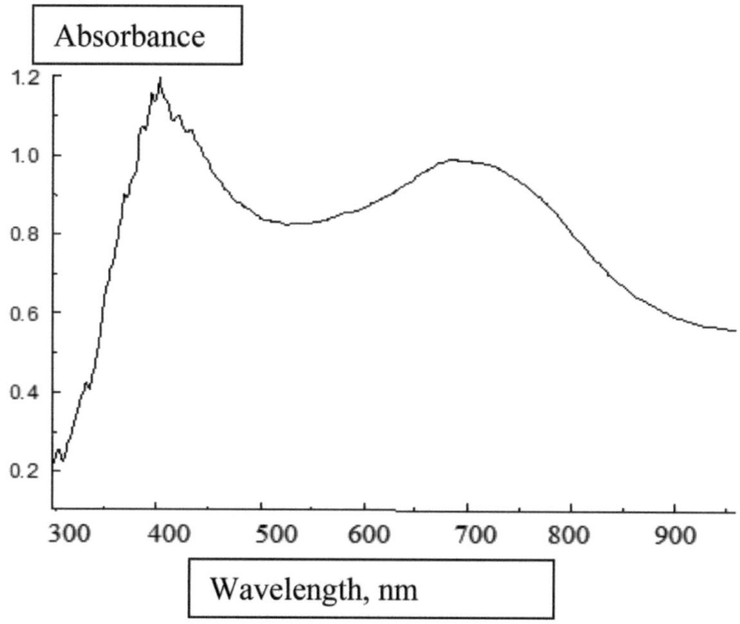

Figure 8. An optical spectrum for the sample of hybrid metal-mesogenic nanosystem Ag/5CB (1:10 w/w %) encapsulated in poly-para-xylylene polymeric film.

The proposed method allows encapsulation from several molecular layers up to several microns. In the optical spectra of such hybrid metal-containing mesogenic systems encapsulated in PPX films at 300 K, the plasmonic bands appear at 420–440 and 500–600 nm, characteristic of the formation of silver nanoparticles and their ordered aggregates, respectively. Increasing the metal concentration led to the formation of anisometric rod-like particles with multimodal plasmonic resonance related to cross and linear resonance. The formation of rod-like metal nanoparticles with anisometric ratio l/d > 10 enhanced plasmonic absorbance in the long-wave spectral range ($\lambda \geq 650$ nm). In this case, the value of dichroic ratio $D_{\parallel}/D_{\perp}$ in the plasmonic band for Ag/5CB (1:10) samples uniformly oriented by magnetic field rises by 5 times in comparison with pure 5CB samples. Optical properties of the samples are stable for several months.

5. Self-organization processes in organosols and cholesteric mesophases of cholesterol- and thiocholesterol-based nanosystems

Hybrid "silver-cholesterol" and "silver-thiocholesterol" nanosystems were produced by cryochemical synthesis via low-temperature co-condensation of components' vapors or two-phase chemical reduction of the silver nitrate salt by $NaBH_4$ and metal-organic complex transferring into toluene environment using inter-phase zwitter-ionic carrier $N^+(C_8H_{17})_4Br^-$ and functionalized cholesteric ligands [21, 22]. Cholesterol (a) and thiocholesterol (b) molecules were also used in order to form protective surface layers on silver nanoparticles formed in the system, their chemical structures are presented below.

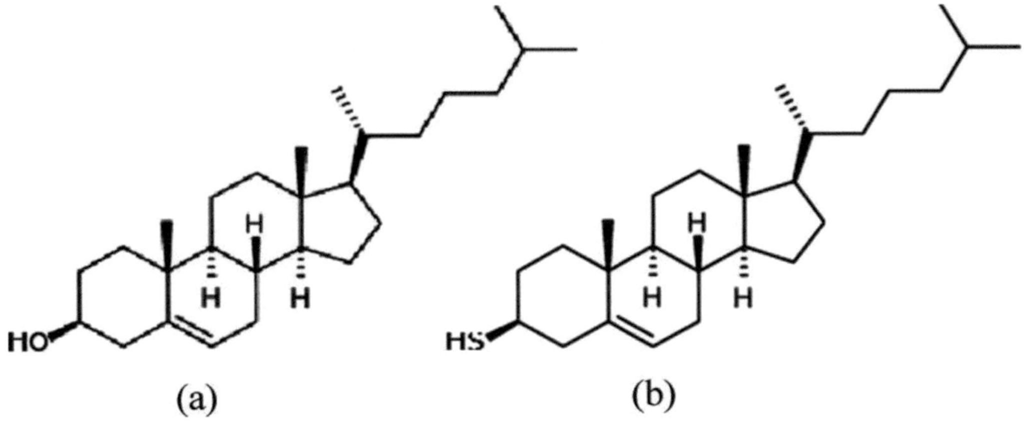

The organic silver nanoparticle's sols prepared after drying and dissolution had been used for FTIR-and UV-Vis spectroscopic studies. The FTIR spectra of the samples were recorded using Tensor II FTIR spectrometer Bruker (Germany), and UV-Vis spectra were recorded by the V770 spectrometer, JASCO (Japan). The chemical composition of the samples obtained was controlled using selected area electron diffraction (SAED) and direct chemical analysis by ICP-MS spectrometer. The internal structure of the samples and the size distribution of the stabilized silver particles were obtained using transmission electron microscopy (TEM) instrument LEO 912 AB Omega (Germany); differential scanning calorimetry (DSC) instrument STA 449C, Netzsch (Germany); and X-ray diffraction instrument D/MAX-2600, Rigaku (Japan).

Understanding the mechanism of interaction between metal species and stabilizing agents—cholesterol and thiocholesterol ligand molecules—was one of the purposes of this study. The most typical bands of O–H group stretching vibrations at $\nu = 3490–3500\ cm^{-1}$ and O–H group δ-deformation vibrations at $\nu = 1090–1140\ cm^{-1}$ and also of S–H group stretching vibrations at $\nu = 2540–2590\ cm^{-1}$ are almost absent in the FTIR spectra of the silver-containing hybrid nanosystem's samples, and the bands stretching vibrations for CH_3–, CH_2–, and S(O)–CH_2 groups at $\nu = 2920–2880\ cm^{-1}$ and $\nu = 1415–1440\ cm^{-1}$ are appear in the spectra. This fact could be considered at the spectral confirmation of the interaction of the thioxy group of thiocholesterol ligand and hydroxy group of cholesterol ligand molecules with the surface atoms of silver nanoparticles [16] and of the existence of stabilizing ligand monolayers on the surface of the metal nanoparticles [17].

The morphology and size distributions of metal species primarily formed in "silver-cholesterol-toluene" and "silver-thiocholesterol-toluene" triple hybrid nanosystems are presented in **Figures 9A** and **B** and **10A** and **B**, respectively. The main diameter values of silver nanoparticles were $d = (6.5 \pm 1.5)$ nm for cholesterol-stabilized organosols of silver nanoparticles and $d = (2.5 \pm 0.5)$ nm for thiocholesterol-stabilized organosols of silver nanoparticles.

The formation of dispersions of stabilized nanoparticles and their regular surface structures took place by withdrawal of organic solvent from the film samples. The chemical nature of stabilizing ligand molecules and concentration of metal nanoparticles in the hybrid system determined the manner of metal species packing in 2D- and 3D superlattices. By removing the

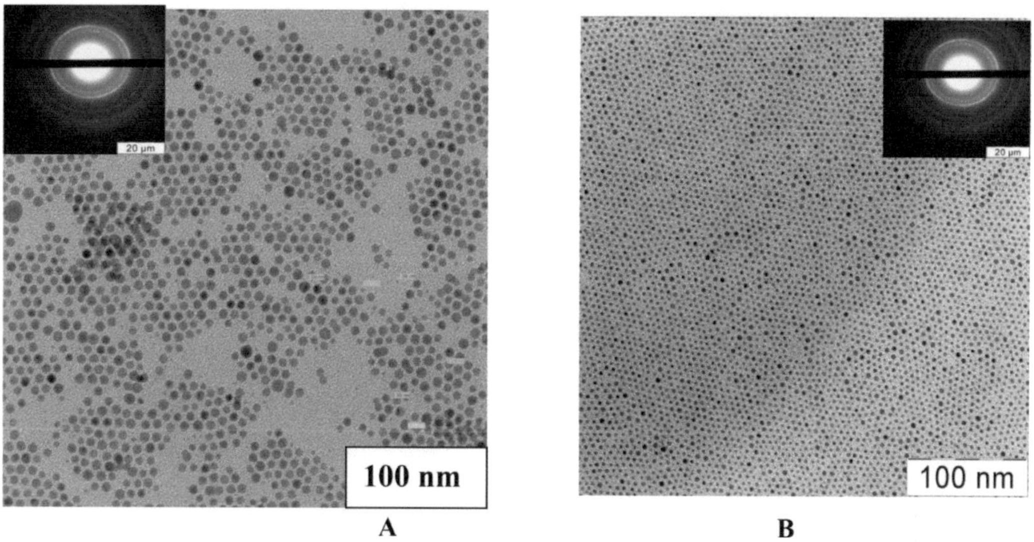

Figure 9. TEM pictures for silver nanoparticles stabilized by cholesterol (A) and thiocholesterol (B) ligand molecules.

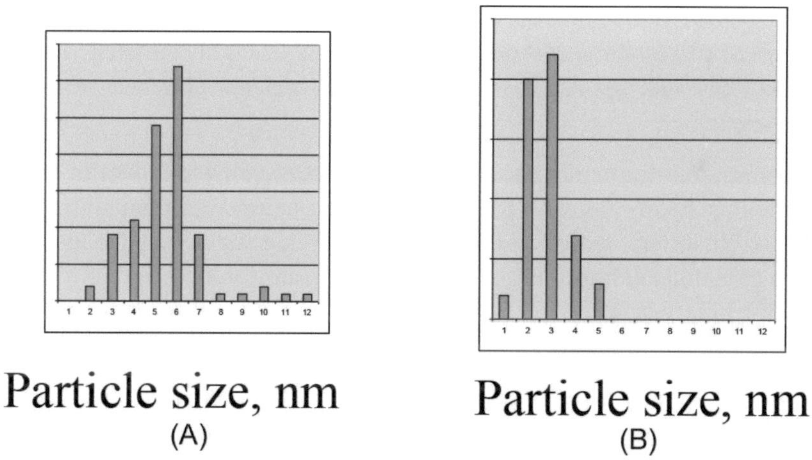

Figure 10. Histograms of size distribution for silver nanoparticles stabilized by cholesterol (A) and thiocholesterol (B) ligand molecules.

organic solvent, the formation of 2D nanostructures with a hexagonal (**Figure 9**) or cubic symmetry was achieved (**Figure 11**).

The process of such ordered nanostructure formation can be described by the following scheme presented in **Figure 12** [20]. It began from the formation of individual metal nanoparticles in highly diluted organic sols stabilized by cholesterol or its heteroatomic derivative thiocholesterol and by raising the metal nanoparticles' concentration, which led to the production of 2D- and 3D nanoparticle superlattices with hexagonal and cubic symmetry.

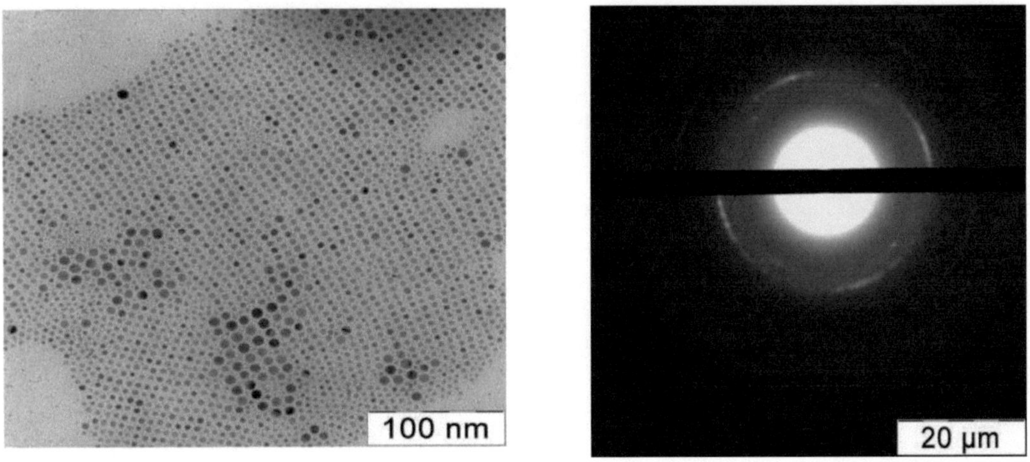

Figure 11. TEM and SAED pictures of silver-thiocholesterol samples obtained by deposition from triple "silver-cholesterol-toluene" system [20].

It is of interest that the interparticle distance (average distance between the centers of metal nanoparticles) in the nanostructures obtained is rather small and is practically equal to the size of organic layer shell from cholesterol or thiocholesterol molecules (about 2.5 nm). Such regular position of plasmonic metal small nanoparticles in highly ordered dielectric organic media promised possible applications of the hybrid systems obtained in nanooptics and molecular electronics.

Binary "silver-thiocholesterol" nanosystem, obtained after entire removal of the toluene solvent, was heated up to 70–80 C and then cooled down to ambient temperature. It possessed liquid crystalline cholesteric mesophase that could exist in a supercooled state for a long time. The formation of elongated nanotubular hybrid aggregates took place. The TEM pictures for the sample of the binary system stored at 293 K for 24 h for the silver nanoparticles formed in the system are presented in **Figure 13**.

Supramolecular ordering of hybrid nanosystem included the interaction of the surface centers of silver nanoparticles with the functional groups of cholesterol and thiocholesterol ligands and their helix supramolecular ordering in cholesteric mesophase. The scheme of molecular interactions is presented in **Figure 14**.

There were intensive plasmonic bands of ligand-stabilized silver nanoparticles at 420–450 nm (**Figure 15**, curve 1) in the UV-Vis spectra of hybrid "silver-cholesterol" and "silver-thiocholesterol" nanosystems. Rising of metal contents led to the red shift of these bands due to specific interactions of silver nanoparticles with ligand molecules and particle-particle dipole-dipole interactions. In the optical spectra of binary "silver-thiocholesterol" system (**Figure 15**, curve 2) along with the plasmonic absorption of individual silver nanoparticles at $\lambda = 450$ nm, the long-wave absorption also appears at 770–1000 nm, which is characteristic for chain aggregates of silver nanoparticles [23]. The hybrid aggregates' absorbance bands were

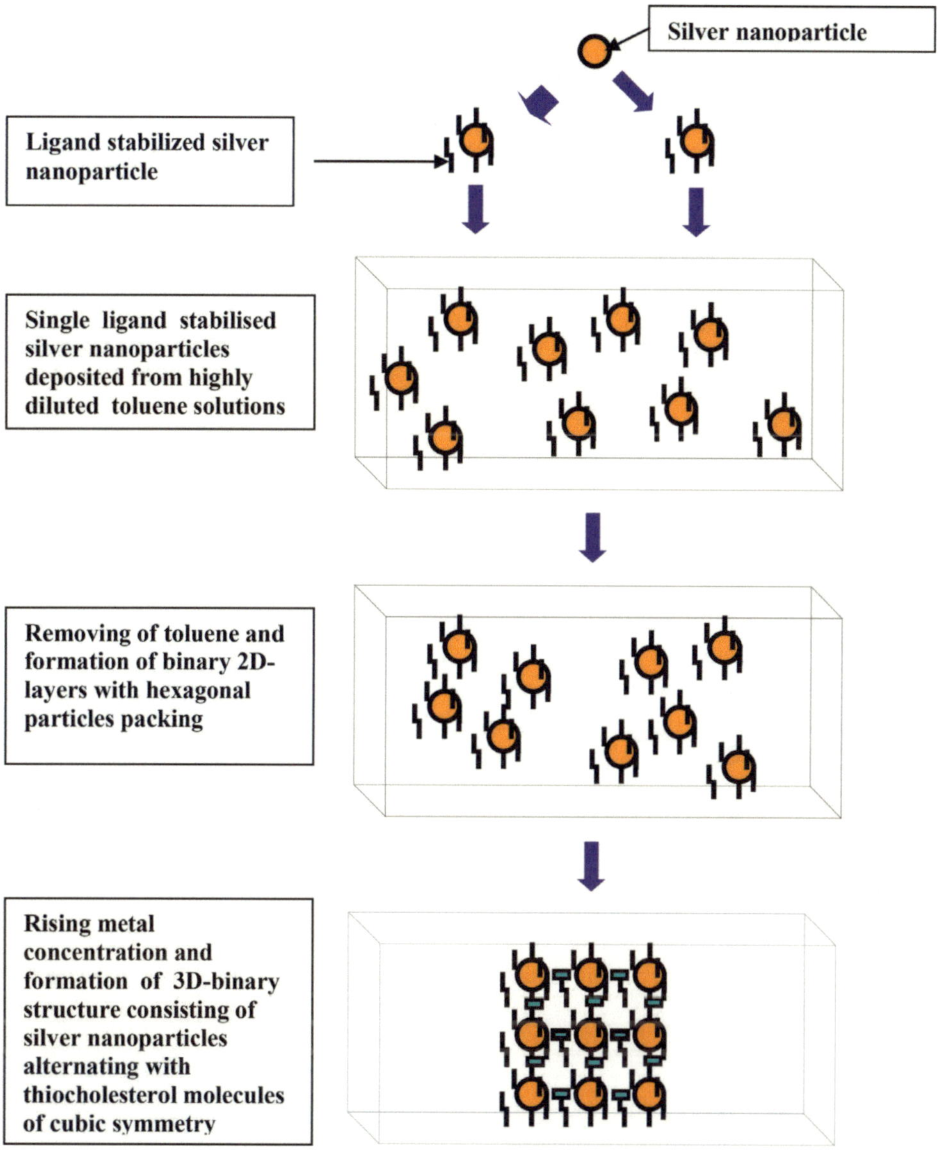

Figure 12. The scheme of formation of binary nanostructures of silver nanoparticles and thiocholesterol molecules in triple "silver-thiocholesterol-toluene" systems [20].

shifted to the near-infrared region and it is important for possible applications in in vivo imaging and sensing of different biomedical objects.

The chirality of active compounds is important for many biochemical processes [24–26], the study and separation of optical isomers of drugs are principal for pharmaceutical applications, and control of optical purity is necessary for the production of pharmaceutical components.

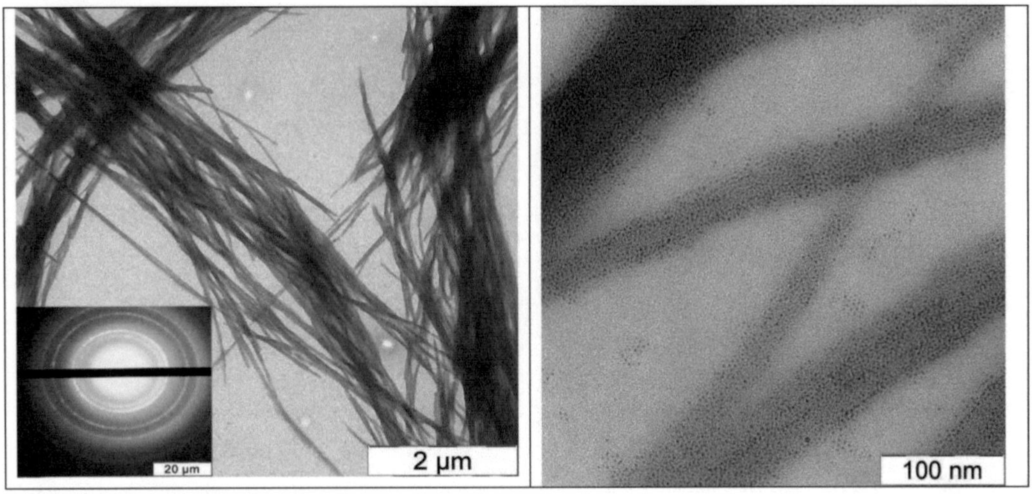

A B

Figure 13. TEM pictures for hybrid silver-thiocholesterol aggregates after 24 h of storage in liquid crystalline mesophase: general view (A) and internal structure (B) [22].

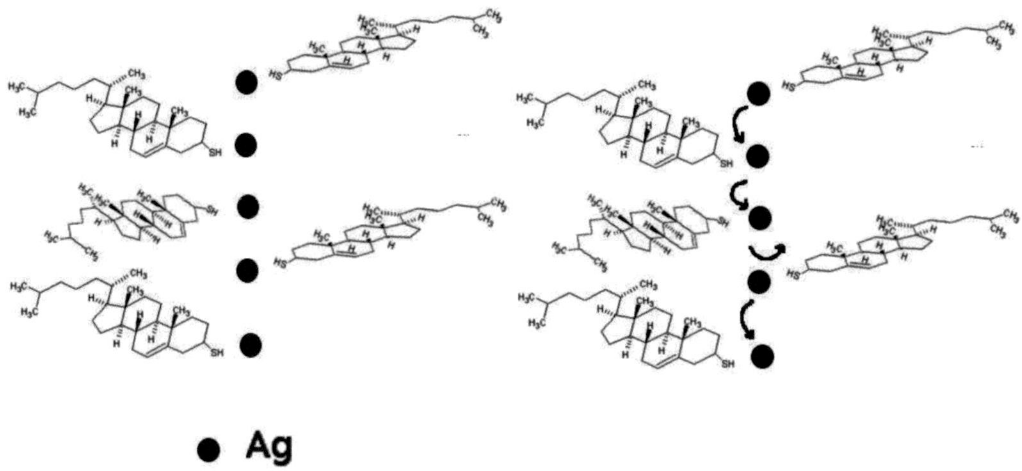

Figure 14. The scheme of silver-thiocholesterol tubular nanoaggregates formation.

Cholesterol and thiocholesterol ligands and their mesogenic derivatives are natural and modified chiral matrices, which adsorptive and selective properties only begin to study [27].

The synthesis of hybrid nanoparticles with silver core and stabilizing cholesterol/thiocholesterol shell layer of chiral ligand thiocholesterol was produced by modified of borohydride reduction method [28], including reduction of silver ions Ag^{+} in bi-phase water-organic system in the presence of mesogenic ligands. This system was used to impregnate silica gel plates for thin-layer chromatography. The UV-Vis spectra of toluene sols of the hybrid

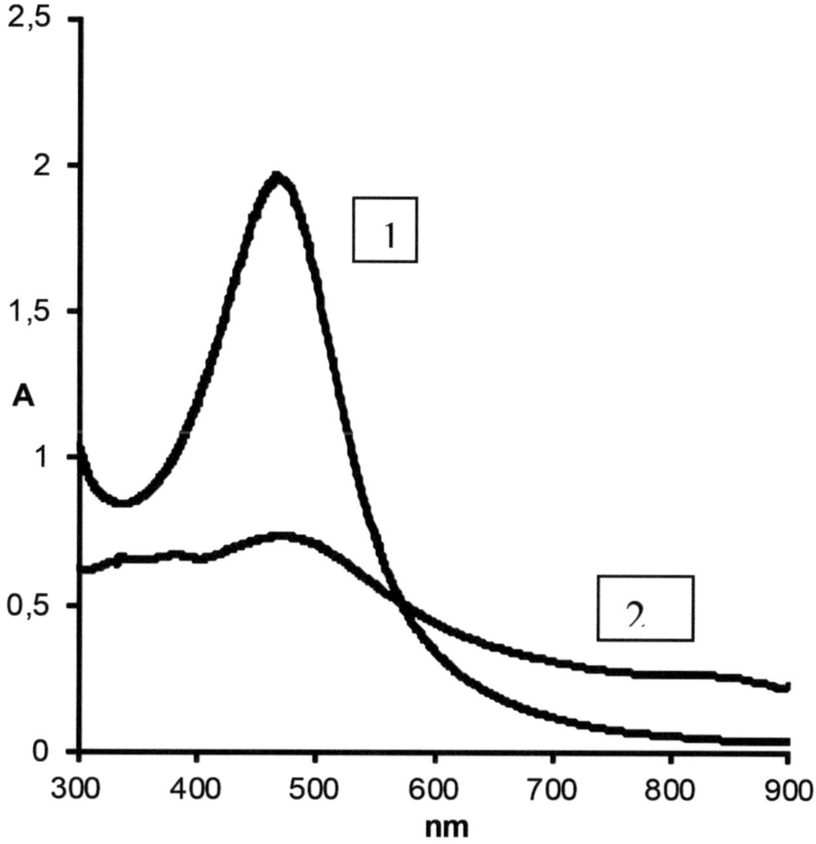

Figure 15. Optical absorbance spectra: 1—highly diluted sols of silver nanoparticles stabilized by thiocholesterol in toluene and 2—linear aggregates, formed in liquid crystalline mesophase of hybrid "silver-thiocholesterol" system.

system "silver-thiocholesterol" (Ag: L = 1:5) were recorded in silver concentration of range 0.2–2.0 mg/ml and show the characteristic silver nanoparticles' plasmonic band at 450 nm. The intensity of absorbance grows linearly by increasing the concentration of nanoparticles.

The selection of model optical isomers of tested compounds was made using silica plates modified by hybrid silver-thiocholesterol (Ag:L = 1:5) nanoparticles (**Figure 16**) with a concentration of 1.2 mg/ml. The chromatographic separation of tested substances was made using modified silica plates [29]. The eluent phase for the selection was chosen from several solvents that differed in their chemical structure and polarity—methanol, isopropanol, acetonitrile, acetone, and butyl acetate. It was shown that the most suitable eluent was the mixture acetonitrile/water (50:50 v/v), the yield mobility was found rising in the order: 1,1′-bi-2-naphtol <2,2′-diamino-1,1′-binaphtol (DABN) ≈ trifluoro antranyl ethanol (TFAE). The difference of the values of R_f was observed for S- and R-isomers of DABN and TFAE. The detection of the spots of separated enantiomers DABN and TFAE was made by UV irradiation at 360 nm of the chromatographic plates using the ability of these compounds for exhibiting fluorescence [30].

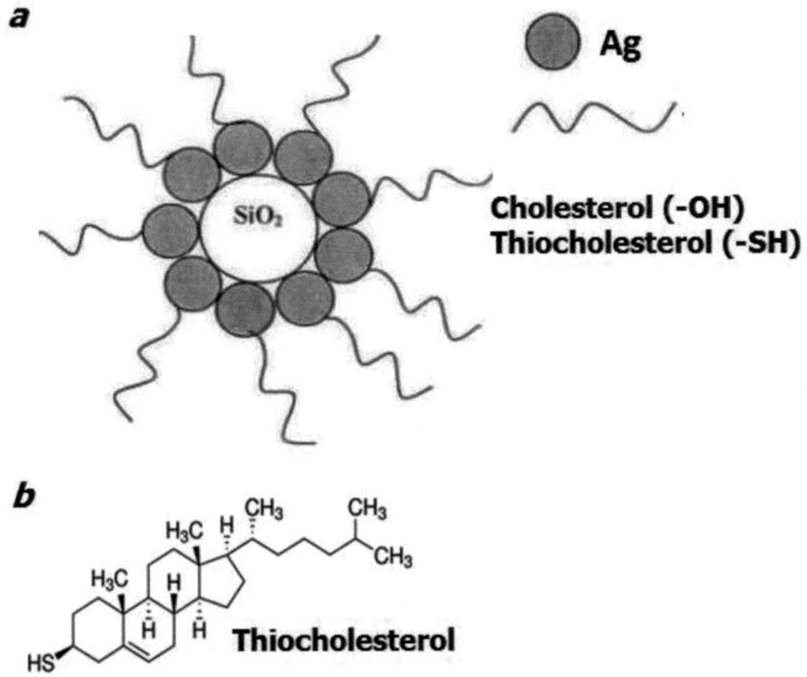

Figure 16. Schematic structure (a) of silica gel modified by silver nanoparticles stabilized by thiocholesterol (b).

The chromatographic separation was estimated by the value of R_f, which was calculated using the relation $R_f = x/L$, where x is the length, passed by the substance from the start line, and L is the length, passed by solvent up to front line [28]. The efficiency of separation was estimated by the value α (selection factor), which was calculated using the relation $\alpha = \frac{1/Rf_2 - 1}{1/Rf_1 - 1}$, where Rf_1 and Rf_2 are parameters of the two substances' retention. The data are presented in **Table 1**. It was shown that hybrid nanosystems based on silver nanoparticles covered by stabilizing layer of mesogenic molecules of thiocholesterol were selective in the separation of optical isomers of DABN and TFAE [31].

Test substances	Rf	α
2,2′-diamino-1,1′-binaphlene (R)	0.57	1.56
2,2′-diamino-1,1′-binaphlene (S)	0.46	
1,1-bi-2-naphthol (R)	0.49	1.07
1,1-bi-2-naphthol (S)	0.51	
TFA (R)	0.43	1.56
TFA (S)	0.54	

Table 1. The retention times (R_f) and separation factors (α) of test compounds in thin-layer chromatography experiments [31].

6. Conclusions and perspectives

A new approach was proposed and realized for the formation of highly ordered hybrid metal-mesogenic nanostructures based on the mesogenic matrix-governed self-organization of atomic-molecular systems obtained by low-temperature condensation of metal and mesogenic components' vapors followed by controlled annealing of the samples. The hybrid metal-mesogenic nanosystems have been obtained including plasmonic metal (silver and copper) nanoparticles stabilized and supramolecular structures organized by liquid crystalline matrix and their properties were studied in temperature range 80–350 K. The conditions that allowed the production of metal nanoparticles of different sizes and morphology were determined, and methods of creation of their ordered nanostructures using molecular self-organization and metal species self-assembly were developed.

Low-temperature co-deposition of metal (Ag, Cu) and mesogenic alkylcyanobiphenyl component (5CB, 8CB) vapors led to the formation of the first stage of metastable biligand complexes of metal atoms with functional CN group, stabilized in ligand matrices. Thermal degradation of the metastable complexes at 150–200 K led to the formation of small metal clusters, which initiate the growth of nanoparticles and their further aggregation and organization in anisotropic mesogenic matrix. The structural and dynamic properties of mesogenic matrix at different temperatures and different phases controlled metal atoms' and small clusters' self-assembling and aggregation and led to the formation of metal particles of different sizes and morphology and their ordered aggregates. The competition of the processes of matrix molecules self-association, metal atoms aggregation and metastable complexes formation between metal species of different size with ligand molecules and molecular aggregates determines the structure and physico-chemical properties of hybrid nanosystems formed in definite conditions. Highly ordered nanohybrid systems can be very promising for different applications in nanoelectronics and electro-optics, as they possess unusual optical and sensor properties.

Ligand-stabilized silver nanoparticles were obtained by biphase chemical synthesis using biomolecule of cholesterol and its heteroatomic analog thiocholesterol. Regular 2D- and 3D superlattices with hexagonal or cubic ordering of almost monodispersed silver nanoparticles with main size d = (6.0 ± 1.5) nm for "silver-cholesterol" and d = (2.5 ± 0.5) nm for "silver-thiocholesterol" systems were obtained by removing the solvent from surface layers of triple "silver-cholesterol (thiocholesterol)-toluene" sols and for binary "silver-thiocholesterol" systems. Hybrid nanosized tubular silver-thiocholesterol aggregates with linear size (length) of more than 10 μm and only 15–20 nm wide including practically monosized silver nanoparticles of d = (2.5 ± 0.5) nm have been formed for binary silver-thiocholesterol system in cholesteric liquid crystalline mesophase.

The optical properties of hybrid nanosystems obtained belong to the plasmon absorption of individual silver nanoparticles and also possess modified plasmonic bands of hybrid nanostructures and highly ordered nanoparticle superlattices formed under special concentration and temperature conditions. Such hybrid nanostructures should be of interest for the creation of new optical and electro-optical devices with adjustable light absorption and light reflection. Integrating silver nanoparticles of exact size and shape and their aggregates of definite morphology with biological molecules and blocks opens new perspectives for obtaining new bio-compatible instruments for imaging and sensing biological objects in living organisms and targeted medical therapy.

The hybrid metal-mesogenic nanosystems obtained were also used as liquid crystalline component for modification of silica gel nanoparticles and production of chiral matrices for thin-film chromatography. The conditions that allowed the production of silica gel plates modified by chiral hybrid metal-mesogenic nanosystems from toluene sols were developed. The new chiral matrices for thin-film chromatography possessed enantioselectivity related to optical isomers of 2,2′-diamino-1,1′-binaphtol (DABN) and trifluoro antrenylethanol (TFAE). We have succeeded in selecting optical isomers of TFAE with selecting factor equal to 1.56. In the future, the hybrid silver-thiochlesterol systems obtained in this work can be used as chiral stationary phase also for high-performance liquid chromatography (HPLC). One can expect the higher values of selectivity effects due to the increasing pathway for selecting compounds in comparison with thin layer chromatography technique.

Acknowledgements

The author would like to thank Dr. S.S. Abramchuk, Centre of Electronic Microscopy of M.V. Lomonosov Moscow State University, for TEM and SAED measurements, and Dr. A. S. Tikhomirov, NT MDT, for AFM measurements.

The author appreciates the help and input of collaborators from Analytical Department of M.V. Lomonosov Moscow State University, and Dr. E.N. Shapovalova, Dr. N.B. Rozhmanova, and Prof. O.A. Shpigun for participation in analytical experiments and fruitful discussions.

The work was financially supported by Russian Foundation of Basic Research (Grant Nos. 15-03-00792-a and 18-03-00730-a).

Author details

Tatyana I. Shabatina

Address all correspondence to: tatyanashabatina@yandex.ru

Department of Chemistry, M.V. Lomonosov Moscow State University, Moscow, Russia

References

[1] Blinov LM. Liquid Crystals: Structure and Properties. Moscow: Publ. Librocom; 2013. 480p

[2] Schmidt G. Nanoparticles from Theory to Applications. Weinheim: Wiley-VCH; 2005. 280p

[3] Klabunde KJ, Richards RM, editors. Nanoscale Materials in Chemistry. 2nd ed. NY: John Wiley & Sons; 2009. 777 p

[4] Sergeev GB, Klabunde KJ. Nanochemistry. NY: Elsevier B.V; 2013. 249 p

[5] Jane PK, Huang X, El-Saed IH, El-Saed MA. Noble metals on the nanoscale: Optical and photothermal properties and some applications in imaging, sensing, biology and medicine. Accounts of Chemical Research. 2008;**41**:1578-1586

[6] Talapin DV, Lee J, Kovalenko MV, Shevchenko EV. Prospects of colloid nanocrystals for electronic and optoelectronic applications. Chemical Reviews. 2010;**110**:389-458

[7] Millstone JE, Hurst SJ, Metraux GS, Cutler JI, Mirkin CA. Colloidal gold and silver triangular nanoprisms. Small. 2009;**5**(6):646-664

[8] Wang L, Xu L, Kuang H, Xu C, Kotov NA. Dynamic nanoparticle assemblies. Accounts of Chemical Research. 2012;**45**(11):1916-1926

[9] Grzekzak M, Vermant J, Furst EM, Liz-Marzan LM. Directed self-assembly of nanoparticles. ACS Nano. 2010;**4**(7):3591-3605

[10] Nie Z, Petukhova A, Kumacheva E. Properties and emerging applications of self-assembled structures made from inorganic nanoparticles. Nature Nanotechnology. **5**:15-25

[11] Macfarlane RJ. Nanoparticle superlattice engineering with DNA. Science. 2010;**201**(334): 204-208

[12] Dong A, Chen J, Vora PM, Kikkawa JM, Murray CB. Binary nanocrystal superlattice membranes self-assembled at the liquid-air interface. Nature. 2010;**446**:474-478

[13] Okuda M. Fe_3O_4 nanoparticles: Protein mediated crystalline superstructures. Nanotechnology. 2012;**23**:415601

[14] Grill L, Dver M, Lafferenz L, Persson M, Peters MV, Hecht S. Nano-architectures by covalent assembly of molecular building blocks. Nature Nanotechnology. 2007;**2**(11):687 691

[15] Stoeva S, Klabunde KJ, Sorensen CM, Dragieva I. Gram-scale synthesis of monodisperse gold colloids by the solvated metal atom dispersion method and digestive ripening and their organization into two- and three dimensional structures. Journal of the American Chemical Society. 2002;**124**(10):2305-2308

[16] Pileni M-P. Self-assembly of inorganic nanocrystals: Fabrication and collective intrinsic properties. Accounts of Chemical Research. 2007;**40**:685-693

[17] Shabatina TI, Sergeev GB. Chapter 11: Cryochemistry of nanometals. In: Hasmin AA, editor. Polymer Thin Films. Vienna: In-Tech Publ; 2010. pp. 185-196

[18] Shabatina TI. Cryoformation of hybrid metal-mesogenic nanosystems. Molecular Crystals and Liquid Crystals. 2011;**545**:1268-1275

[19] Shabatina TI, Belyaev AA, Sergeev GB. Self-assembled nanostructures in silver/thiocholesterol and silver/cholesterol systems. BioNanoScience. 2013;**5**:284-292

[20] Shabatina TI, Belyaev AA, Morosov YN. Self-organisation processes and phase-structural states in hybrid metal-mesogenic nanosystems. Molecular Crystals and Liquid Crystals (Taylor & Francis, UK). 2016;**688**:55-64

[21] Shabatina TI, Belyaev AA, Sergeev GB. Silver/thiocholesterol and silver/cholesterol nano-sized aggregates formation in liquid crystalline mesophase. Molecular Crystals and Liquid Crystals. 2011;**540**:169-174

[22] Belyaev AA, Shabatina TI, Sergeev GB. Nanosized silver/thiocholesterol aggregates. Moscow University Chemistry Bulletin. 2009;**64**(4):182-185

[23] Karpov SV, Slabko VV. Optical and Physical Properties of Fractal Structured Metal Sols. Novosibirsk; 2003. 198 p

[24] Allenmark SG. Chromatographic Determination of Enanthiomers. Nauka; 1991. 268 p

[25] Ahuja S, Allenmark SG, Krstulovic AM, Souter RW, Stevenson D. Chiral separation by liquid chromatography. Journal of Chromatography. 1998;**6**:641

[26] Bhushan R, Ali I. Chromatographic separations of steriomers. Chromatographia. 1993;**35**: 679

[27] Shapovalova EN, Anan'eva IA, Mazhuga AG, Zik NV, Shpigun OA. Application of gold nanoparticles for immobilisation of L-cystein on plates for thin layer chromatography. Moscow University Bulletin. 2009;**5**:200-207

[28] Pesek JJ, Matyska MT, Dawson GB, Wilsdorf A, Marc P, Padki M. Cholesterol bonded phase as a separation medium in liquid chromatography. Journal of Chromatography. A. 2003;**986**:253-262

[29] Gromova YA, Anistratova ES, Shabatina TI, Rozhmanova NB, Shapovalova EN, Shpigun OA. Chiral matrices production based on hybrid nanosystems Silver-ThioCholesterol and study of their properties. Moscow University Bulletin: Series 2. 2016;**57**(1):35-39

[30] Lisichkin GV. Modified Silica in Adsorption, Catalysis and Chromatography. Moscow: Chemistry; 1986. 286 p

[31] Shabatina TI, Gromova YA, Anistratova ES, Belyaev AA. New chiral metal-mesogenic nanosystems "silver-thiocholesterol" and their adsorbtion properties. Molecular Crystals and Liquid Crystals (Taylor & Francis, UK). 2016;**632**:64-69